Set Theory and Infinity
An Informal Introduction

Steven Brawer

January, 2022
First Edition
ISBN: 978-0-578-34990-9

Contents

1 introduction

A set is a collection of things. An infinite set is a collection of an infinite number of things. Set theory is, among other things, a mathematical theory of infinity based on infinite sets, and gives precise meaning to the otherwise vague concept of infinity .

This book presents an informal exposition of set theory. The goal of this book is to develop intuition for as well as mathematical fluency in understanding and manipulating sets, particularly infinite sets.

A great deal of mathematics explicitly requires set theory, and even advanced topics are typically formulated in the kind of informal set theory presented here. This book presents the set theory required for such topics as metric spaces, vector spaces, topology, abstract algebra and analysis, on a first or second year graduate level.

This approach is generally simpler, more intuitive and more accessible (and less powerful and less inclusive) than that of **formal set theory.** Formal set theory is based on the ZF (Zermelo-Frankel) or ZFC (C = axiom of choice) axioms, which are not discussed in this book. Since set theory is a branch of mathematical logic, a deep and profound understanding requires the formalism. Nonetheless, this book should be useful. It is likely to be a good idea to have an informal introduction before tackling the formalism.

This book should be accessible to someone with a good grasp of calculus.

2 notation

$\mathbb{Z}$ - set of all integers
$\mathbb{Z}^*$ - set of all integers excluding 0
$\mathbb{Z}_+, \mathbb{N}$- set of positive integers (natural numbers) (excludes 0)
$\mathbb{Q}, \mathbb{Q}^*, \mathbb{Q}_+$ - set of rational numbers
$\mathbb{R}, \mathbb{R}^*, \mathbb{R}_+$ - set of real numbers
ω - $\mathbb{N}$, symbol for the ordinal
$\{1, 2, a\}$ - set of elements 1,2,a
$a \in A$ - a is in, an element of, a member of set A
$a \notin A$ - a is not in, not an element of, not a member of set A
$\emptyset$ - empty set
$\{\}$ - empty set
$\mathcal{P}(A)$ - power set of set A
$|A|$ - cardinality (size) of set A
$\aleph_0$ - cardinality of countably infinite set
$\aleph_1$ - cardinality of uncountable set
$B \subset A$ - set B is a proper subset of set A
$B = A$ - set B equals set A
$A \neq B$ set A not equal to set B
$A \cup B$ - union of sets A, B
$A \cap B$ - intersection of sets A, B
$A/B,\ A - B$ - complement of set B relative to set A
B^c - complement of B relative to A when A is implicit
$\bigcup S$ - union of all sets in S set-of-sets
$\bigcap S$ - intersection of all sets in S set-of-sets
$B \subseteq A$ - set B is either equal to set A or is a proper subset of set A
$A \supset B$ - A is a proper superset of B ($= B \subset A$)
(a, b) - ordered pair
$(a, b), [a, b], (a, b], [a, b]$ - intervals of real numbers a, b
$((a, b)), [[a, b]], ((a, b]], [[a, b))$ - intervals of rational numbers a, b (only rationals in intervals)
$\sim$ - equivalence relation
$a \sim b$ - a equivalent to b
$A \cong B$ - set A isomorphic to set B
$[a]$ - equivalence class of element a
$A/\sim$ - quotient set of set A
$\times$ - cartesian product
$A \times B$ - cartesian product of sets A, B
$a|b$ - integer a divides integer b
$f : A \to B$ - map (or function) from set A to set B
f^{-1} - pre-image or inverse of function f
$F = G \cdot H$ - composition of functions G and H to give function F

$\prec$, $\preceq$ - order relations for cardinal numbers.
T, F - True, False
$\wedge$ - and
$\vee$ - or
$\neg$ - not
$\rightarrow$ - conditional (conditional implies)
$:=$ - assignment for logical expression
$\Rightarrow$ - implies
$\exists$ - existential quantifier
$\forall$ - universal quantifier
$\cong_o$ - order isomorphism

3 very simple description of a set

This section introduces the concept of sets and examples of simple sets. The following section describes some operations on sets, and the section following that describes procedures for creating arbitrary sets.

3.1 sets of integers

A set is a collection of things. It is like a bag in which things can be put, except the bag has no fixed size, may be empty, and is never full. Here is the notation for a set of three integers.

$$S = \{3, 18, 2\} \tag{3.1}$$

S is a variable to which the set is assigned. It can be thought of as giving a reference label to the set. The same set may be assigned to multiple variables. For example

$$W = \{3, 18, 2\} \tag{3.2}$$

The sets S and W above contain (have) three integers: 3, 18, 2. Anything between the curly brackets { and } is an **element** or **member** of the set. So 3 is an *element* of S and of W, 18 is a *member* of S and W.

Here is another set of numbers

$$\{1.34, 9.67, 19034.5643, 000.76, 54\} \tag{3.3}$$

Any numbers can be put into a set in any order. Here is a set of 6 integers.

$$V = \{4, 54, 6, 1, 29038847, 0\} \tag{3.4}$$

V is a weird-looking set, but it is a set nonetheless.

The elements in a set are unordered. Thus we could have written equation 3.1 as

$$S = \{18, 2, 3\}, \quad S = \{2, 3, 18\} \tag{3.5}$$

as well. These are all the same exact set.

> We will usually display sets of numbers in alphanumeric order to make them easier to read. However, the only thing which is important is the contents of the set. The order in which the objects of the set appear is for display purposes only and has no mathematical implications.

The elements of a set must be unique. That is, there cannot be repeats. The set $\{2,4,6\}$ is exactly the same as the set $\{2,4,6,2\}$, and in fact this latter notation is redundant and should not be used.

The notation to indicate that some entity is in a set is to use the symbol $\in$ for **being in** (being an element/member of) and $\notin$ for **not being in** (not being an element/member of). Thus referencing equations 3.1 and 3.4

$$3 \in S, \quad 6 \in V, \quad 54 \notin S, \quad 29038847 \in V, \quad 18 \notin V, \quad 54 \in V \tag{3.6}$$

A **singleton set** is a set with only one element. The set $Y = \{5\}$ is a singleton set.

Definition 3.1

Two sets are ***equal*** *when they have the same elements, regardless of order. If sets A, B are equal, we write $A = B$.*

Two sets are ***unequal*** *when one of the sets contains at least one element which is not in the other. If sets A, B are unequal, we write $A \neq B$*

Two sets are ***disjoint*** *if they have no elements in common. Any number of sets is* ***pairwise disjoint*** *if each pair of sets is disjoint.*

From equations 3.1 and 3.2, $S = W$. Also

$$\begin{aligned} &\{1,2,3,4\} = \{2,4,3,1\} \\ &\{1,2,3,4\} \neq \{1,2,3,4,5\} \\ &A = \{1,2\}, \ and \ \ B = \{2,3\} \Longrightarrow A \neq B \end{aligned} \tag{3.7}$$

The sets $\{1,2,3\}$, $\{4,5,6\}$ are **disjoint**. The sets $\{1,2,3\}$, $\{4,5,6\}$, $\{7,8,9\}$, $\{10,11,12\}$ are pairwise disjoint.

3.2 sets with variables

A set may contain **variables**. For example, let $x = 7$ and $S = \{1,3,5,x\}$. Then $S = \{1,3,5,7\}$. If $x = 2$, $y = 4$, $z = 6$, then $\{x,y,z\} = \{2,4,6\}$.

Let $x = 7, y = 140, z = 0, p = 140$. Then 140 is a repeated element of $\{x,y,z,p\}$.

3.3 sets with abstract elements

It is useful to define sets with abstract elements - that is, elements which are only themselves and don't stand for, or represent, anything else.

Consider the set

$$A = \{a, b, c\} \tag{3.8}$$

What, you ask, are a,b,c? They could be variables but also they could be just themselves - they simply **are** the elements of the set A. They don't necessarily stand for anything. Their most important property is that they are distinguishable one from the other and from numbers. Remember, sets are not allowed repeats, so that the set $\{a, b, a\}$ is **not** a set, because 'a' appears twice. a,b,c are considered to be distinct entities.

We may define

$$a = 5, b = 20, c = 3, \quad G = \{a, b, c\} \tag{3.9}$$

so G is a set of integers, $G = \{3, 5, 20\}$. Now a,b,c are no longer abstract elements but are **variables**, which in this particular case stand for (or represent) integers. It should be obvious from context what is going on.

> In this book, if an alphanumeric element of a set is not a number, then unless it is explicitly stated that the element is a variable, it should be considered an abstract element.

As with integers, order is irrelevant. If a,b,c are abstract elements

$$\{a, b, c\} = \{b, c, a\}$$

We can combine everything introduced so far

$$x = 5, a = 20, z = 15$$
$$\{1, a, 50, x, b, c, z\} = \{1, 5, 15, 20, 50, b, c\}$$

Note that the right hand side of the second expression is written in alphanumeric order, which is useful for display purposes.

For those with a background in computer programming, one can think of an abstract member of a set as a pointer to a unique object which has no interpretation, no usefulness and no operations. Different abstract elements point to distinct objects (distinct memory locations).

3.4 definition of a set

Here is an **attempt** to define a set.

> **Definition 3.2**
>
> *A set is is a collection of elements.*

This statement is essentially meaningless because the terms *collection* and *element* must be defined. **The concept of a set is intuitive.** The consensus among mathematicians seems to be that it is not possible to define the concept of set with rigor. The concept is understood by metaphor. For example, a set is like a bag into which things can be put. Unlike a bag, there is no such thing as a full set. Elements can be added to any set. This is done with an operation called a *union*, which is introduced later.

In formal set theory, a set is something which obeys the formal axioms of ZF or ZFC, and nothing else. But we are not going to be formal, so we need an intuitive and useful and sensible idea of what a set is.

The terms **collection, family, class** are common synonyms for *set*. Sometimes they are used in particular situations, which we will discuss later.

3.5 the empty set

A set with no elements is called the **empty set**. An empty set is just as real and specific and important as sets with elements, such as S, T, V above. The following two expressions are the same, and each assigns the empty set to the variable E.

$$E = \emptyset \;\;, \quad E = \{\} \tag{3.10}$$

Symbolically the empty set is represented either by a pair of brackets $\{\}$ containing nothing, or by the symbol $\emptyset$, which is a letter in the Norwegian and Danish alphabets (it is not the same thing as 0 - zero).

As will become clear, the empty set is a strange creature and is very useful.

3.6 cardinality of a set = size of the set

Definition 3.3

*A **finite set** has a finite number of elements. A set which is not finite is called **infinite**.*

*The **size** of a finite set is the number of elements in the set. The size is called the **cardinal number**, or **cardinality**, of the set.*

The sets S, V above, equations 3.1, 3.4, are finite. The empty set is finite (Since the empty set has 0 elements, and 0 is a finite number, the empty set is finite.). The 'size' of an infinite set is defined later.

The size of the set S, equation 3.1, is 3, and the size of set V, equation 3.4, is 6. Symbolically we write the size of a set as |S|, |V|, and so forth. Thus we have

$$|S| = 3, \ \ |V| = 6 \tag{3.11}$$

Definition 3.4

*In set theory, the size of a set is called the **cardinality** of the set. The cardinality of a set S is notated |S|. If a set S is finite, |S| is an ordinary integer. When this integer refers to the size of a set, it is called a **cardinal number**.*

The following, regarding infinite sets, is likely to be confusing at this point but will become clear. The term *cardinality* applies to both finite and infinite sets. The cardinality of an infinite set is very different from that of a finite set. An infinite set does not have a concept of 'number of elements'. The cardinality of an infinite set is not an ordinary integer. There are symbols which represent the concept of the 'size' of an infinite set just as there are symbols (1,2,...) representing the size of finite sets. The symbol *is* is the cardinal number, or cardinality, of that set. As will be seen, the cardinality (size) of a finite set may be defined in such a way that the definition can be generalized to define the the cardinality of an infinite set. Later the addition of cardinalities of finite and infinite sets will be defined. This will give an arithmetic of cardinal numbers involving these new symbols.

In this book, the word **cardinality** is preferred over the word 'size'.

The cardinality of the *empty set* is 0. Using the notation of equation 3.11 for the cardinality of a set

$$|\emptyset| = |\{\}| = |E| = 0 \tag{3.12}$$

where E is as defined in equation 3.10.

3.7 sets with big integers

Integers can be very large. For example

$$x = 3781992038986753647812937 \tag{3.13}$$

is an integer. Most people are not used to a number this big. After all, what would it be used for unless, perhaps, to specify the size of some astronomical entity in centimeters with considerable precision. Still, x is a genuine integer. While x seems like a large integer, it is in fact 50 orders of magnitude smaller than the integer

$$y = 3781992038986753647812937 \times 10^{50} \tag{3.14}$$

which is 100 orders of magnitude smaller than the integer

$$z = 3781992038986753647812937 \times 10^{150} \tag{3.15}$$

and in fact these numbers aren't even close to infinity (whatever that is). Anyway, the following set is a perfectly valid set of three distinct integers

$$\begin{aligned} \{x, y, z\} = \{&3781992038986753647812937, 3781992038986753647812937 \times 10^{50}, \\ &3781992038986753647812937 \times 10^{150}\} \end{aligned} \tag{3.16}$$

4 some operations on sets

This section presents two basic operations on sets: union and intersection. There exist more operations than are presented here. These will be presented later.

Set operations will be illustrated using the following sets:

$$\begin{array}{lll} A = \{1,3,4,7,10\} & B = \{1,4,8,10,12,15,21,40\} & \\ C = \{7,9,12,100\}, & ,D = \{21,41,92\}, & E = \emptyset \\ F = \{1,3,4,7,10,12,15\}, & G = \{1,3,4,7,10\} & \end{array} \tag{4.1}$$

Note that $A = G$.

4.1 union

4.1.1 union of two sets

Definition 4.1

The ***union*** *of two sets is a set containing all the elements of both sets, with no repeats.*

Given arbitrary sets P, Q which may be empty, finite or infinite. The union of sets P, Q is notated with the ***cup*** *symbol $\cup$.*

$$R = P \cup Q = Q \cup P \tag{4.2}$$

where R is a the set which results from the union. R is a set with all the elements of P and Q, with no repeats. If P, Q are disjoint, $R = \emptyset$.

The order does not matter (the operation is associative).

The result of a union of two sets is a third set. This third set contains all the elements which are in one of the original sets, all the elements which are in the other original set, and all the elements which are in both, but with **no repeats**.

Here are examples using sets from equation 4.1.

$$Q = A \cup B = \{1,3,4,7,8,10,12,15,21,40\} \tag{4.3}$$

The set which is the result of the union is Q. Note that the elements 1, 4, 10 appear only once in the set Q, even though these elements are found in both sets A, B. Any element in Q is either in A, in B, or in both.

Note the cardinalities (sizes).The cardinality of a finite set is an ordinary integer.
$|Q| = 11$, while $|A| = 5$, $|B| = 8$, so that $|A| + |B| = 13 > |Q| = 11$.

Here are two other examples:

$$A \cup F = \{1, 3, 4, 7, 10, 12, 15\}, \qquad A \cup G = \{1, 3, 4, 7, 10\} = A = G \tag{4.4}$$

Note the cardinalities.

$$\begin{aligned} |A| + |F| = 12 > |A \cup F| = 7 \\ |A| + |G| = 10 > |A \cup G| = 5 \end{aligned} \tag{4.5}$$

Since the empty set has no elements, the union $X \cup \emptyset = \emptyset \cup X = X$ for **any** set X. The empty set contains no elements to add to a union. There are three ways to write the **union with an empty set**:

$$\begin{aligned} C \cup E = \{7, 9, 12, 100\} \\ C \cup \emptyset = \{7, 9, 12, 100\} \\ C \cup \{\} = \{7, 9, 12, 100\} \end{aligned} \tag{4.6}$$

where E is defined as the empty set in equation 4.1.The above three equations are equivalent.

The union of two empty sets is the empty set.

$$\{\} \cup \{\} = \{\} \cup \emptyset = \emptyset \cup E = ... = \emptyset \tag{4.7}$$

Other examples from equations 4.1:

$$\begin{aligned} C \cup E = \{7, 9, 12, 100\} = C \\ C \cup D = \{7, 9, 12, 21, 41, 92, 100\} \end{aligned} \tag{4.8}$$

Here is an algorithm to create a union of two sets.

Definition 4.2

Algorithm for $R = P \cup Q$. Start off with R as the empty set, $R = \emptyset$. Copy all elements of P into R. Then copy all elements of Q into R, except for duplicates. The result set R is the union.

This is valid even if one or both sets are infinite, even though the 'copying' requires an infinite number of steps. The idea of an apparently infinite number of iterations is discussed later.

Now we have a very important theorem about cardinalities of a union of two *finite* sets.

Theorem 4.3

It is always the case that, for any two ***finite*** *sets A, B*

$$|A| + |B| \geq |A \cup B| \tag{4.9}$$

Only when A and B are disjoint (no elements in common) does $|A|+|B| = |A \cup B|$.

The $>$ (strictly greater than) of this theorem **must** apply when A and B contain at least one element in common. As an exercise, prove (or at least rationalize) this theorem.

Here is an example of the disjoint case: $P = \{1, 2, 3\}, Q = \{a, b, c, d\}$, $P \cup Q = \{1, 2, 3, a, b, c, d\}, 3 + 4 = |P| + |Q| = |P \cup Q| = 7$.

4.1.2 union of multiple sets

The union of multiple sets is a straightforward generalization of the union of two sets. The resultant set contains all the elements of all the sets, with no repeats. For example, for sets of equation 4.1

$$A \cup B \cup C = \{1, 3, 4, 7, 8, 9, 10, 12, 15, 21, 40, 100\} \tag{4.10}$$

Equation 4.9, for cardinality, generalizes.

$$17 = 5 + 8 + 4 = |A| + |B| + |C| > |A \cup B \cup C| = 12 \tag{4.11}$$

The algorithm, definition 4.2, generalizes for $R = A \cup B \cup C$. Start off with $R = \emptyset$. Add all elements of A to R. Then add all elements of B, without duplicates. Then add all elements of C, without duplicates. For $R = A \cup B \cup C$, if A, B, C *pairwise disjoint* (so no set contains any element of any of the other sets), the generalization of the second part of theorem 4.3 holds, so $|R| = |A| + |B| + |C|$.

The algorithm, definition 4.2, is generalized straightforwardly for an infinite number of sets, each of which can be finite or infinite. A notation for these situations is given later.

Parentheses may used to group expressions within a larger expression. Parentheses work for sets a similar way that they work for arithmetic. For example 2+3+7 = (2+3)+7=2+(3+7)

Like addition, a string of unions can occur in any order.

$$R = A \cup B \cup C = (A \cup B) \cup C = A \cup (B \cup C) \tag{4.12}$$

In the middle expression, first perform a union of A and B, and then perform the union of the resulting set with C. The third expression means first union B and C, then union the resulting set with A. That is

$$\begin{aligned} Q &= A \cup B \\ R &= Q \cup C \\ T &= B \cup C \\ R &= A \cup T \end{aligned} \tag{4.13}$$

This generalizes in an obvious way to an arbitrary but finite number of unions. Generally a different notation is used for infinite unions.

4.1.3 insertion into a set

We can use the union to define the operation of **insertion into a set**. Let $A = \{1, 2, 3, 4\}$. Consider the following sequence of operations.

$$\begin{aligned} A &= \{1, 2, 3, 4\} \\ B &= A \cup \{5\} \\ A &= B \end{aligned} \tag{4.14}$$

The result is that $A = \{1, 2, 3, 4, 5\}$. We have **inserted** 5 into the original set A, called the result B, then renamed the result A. We might as well skip the intermediate set B.

Definition 4.4

*Given set P, the **insertion** of the contents of set Q into set P is notated*

$$P = P \cup Q \tag{4.15}$$

4.2 diagrams of sets

A **diagram of a set** is a way of visualizing sets and operations on sets. There are a number of different types of diagrams. Here we present one type, useful for finite sets. For example, to visually represent the set $A = \{2, 5, 12, 20\}$ we can draw diagrams like the following:

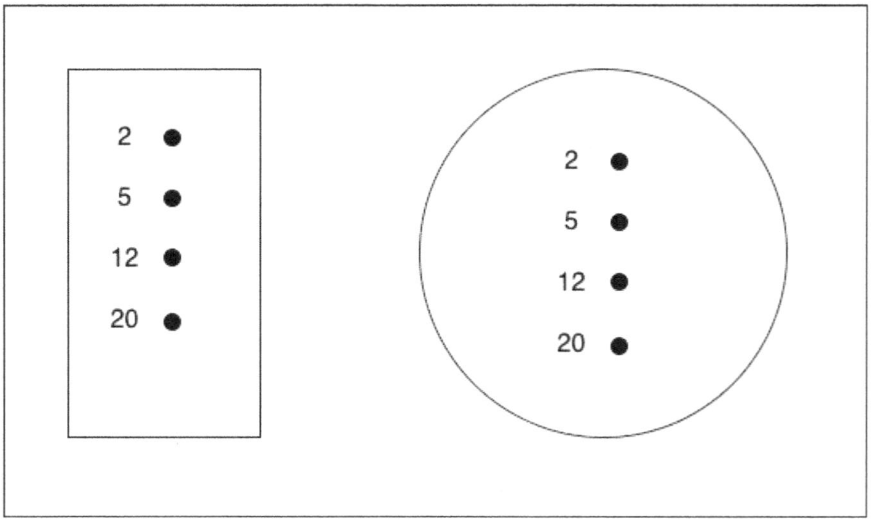

Figure 4.1:

We have some closed shape - a circle, square, rectangle, a closed irregular shape - in which there is a label for each element. We may (but also may not) place dots representing the elements of the set next to the labels, as in the above figure.

Diagrams can represent relationships between two or more sets. Below is a diagram representing the sets

$$A = \{2, 4, 6, 8\}, \quad B = \{6, 8, 10\} \tag{4.16}$$

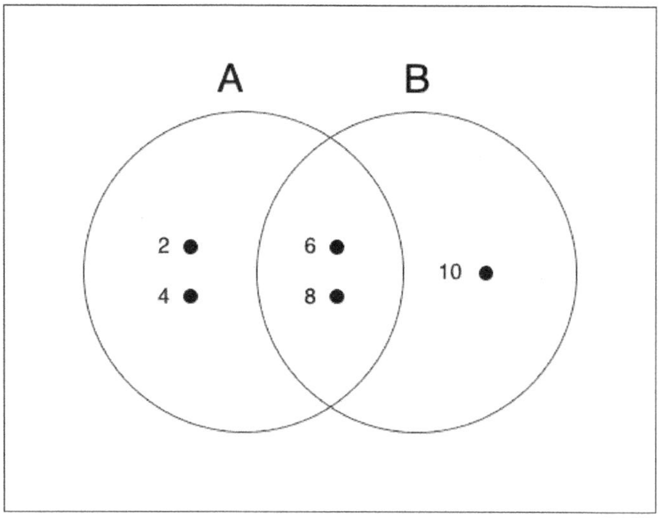

Figure 4.2:

Thew numbers 2, 4 are in A only, 10 is in B only and 6, 8 are in both A and B.

4.3 intersection

In figure 4.2 above, the numbers 6, 8 are common to sets A and B. We say that the **intersection** of sets A and B equals the set $\{6, 8\}$.

Definition 4.5

Given any two sets P, Q, the ***intersection*** *of the two sets, notated*

$$R = P \cap Q = Q \cap P \tag{4.17}$$

is a third (new) set R containing all the elements ***common*** *to both P and Q. The operation is associative.*

The intersection of sets A, B, equation 4.16 is the set $\{6, 8\}$

$$\{6, 8\} = A \cap B = \{2, 4, 6, 8\} \cap \{6, 8, 10\} \tag{4.18}$$

The intersection is represented by the **overlap area** of figure 4.2. **Intersections are visually illustrated as the overlap area of diagrams.**

Let $C = \{a, b, c, d\}$. Some additional examples:

$$A \cap C = \emptyset \quad A \cap A = A \tag{4.19}$$

Here is an algorithm to create an intersection of two sets.

Definition 4.6

Algorithm for $R = P \cap Q$. *Let* $Z = P \cup Q$ *(so* Z *contains all elements of* P *and* Q*). Start off with* $R = Z$. *Then intersect* P *with* R*, so* $R \to R \cap P$. *Then intersect* Q *with* R*, so* $R \to R \cap Q$. *The result set* R *is the intersection.*

In other words, $P \cap Q = (P \cup Q) \cap P \cap Q$. (Parentheses are important here.) For example, Let $P = \{1, 7, 10\}, Q = \{7, 18, 10\}, P \cup Q = \{1, 7, 18, 10\}$.

$$P \cap (P \cup Q) = \{1, 7, 10\} = P$$
$$Q \cap (P \cap (P \cup Q)) = Q \cap P = \{7, 10\}$$

Exercise: validate the algorithm.

Several intersections of the sets $V = \{1, 2, 3, 4, 5\}$, $W = \{3, 4, 5, 6, 7\}$ and $X = \{1, 4, 7, 12\}$ are

$$V \cap W \cap X = \{4\}, \quad W \cap X = \{4, 7\}, \quad V \cap W = \{3, 4, 5\}, \quad V \cap X = \{1, 4\} \tag{4.20}$$

and are illustrated by the diagram below.

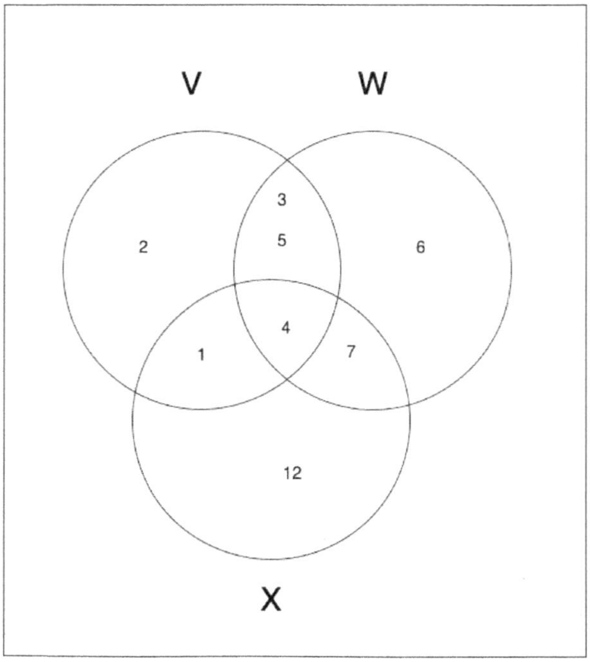

Figure 4.3:

Like unions, intersections can occur in any order, with parentheses grouping pairs of intersections.

$$A \cap B \cap C = (A \cap B) \cap C = A \cap (B \cap C) \tag{4.21}$$

Disjoint sets have no elements in common, as defined in definition 3.1. If sets A, B are disjoint, then $A \cap B = \emptyset$.

It should be obvious that

$$\emptyset \cap \emptyset = \emptyset \tag{4.22}$$

If Q is any nonempty set

$$\emptyset \cap Q = Q \cap \emptyset = \emptyset \tag{4.23}$$

because Q and $\emptyset$ have no elements in common.

Parentheses are usually required for expressions which mix intersection and union, just as when addition and multiplication are mixed. Thus

$$18 = (3 \times 4) + 6 \neq 3 \times (4 + 6) = 30 \tag{4.24}$$

Analogously

$$\begin{aligned} &A = \{1, 2, 3\}, \quad B = \{a, b, c\}, \quad C = \emptyset \\ &\emptyset = C \cap (A \cup B) \neq (C \cap A) \cup B = \emptyset \cup B = \{a, b, c\} \end{aligned} \tag{4.25}$$

4.4 subset, superset

In this section, we will see that interpreting apparently simple definitions or theorems can sometimes be quite tricky. One must pay careful attention to the wording. This can be rather like legal analysis or hairsplitting.

Here's the definition of *subset*. The definition has several parts, and it is convenient here to give it all at once. We will unpack it, line by line, below.

Definition 4.7

For any two sets P, Q, finite or infinite:

If all the elements of P are in Q we say that P is a ***subset*** *of Q and write*

$$P \subseteq Q \tag{4.26}$$

If P and Q have exactly the same elements, then we say P ***equals*** *Q and write*

$$P = Q \tag{4.27}$$

If Q has at least one element which is not in P, and in addition all elements of P are in Q, then P is a ***proper subset*** *of Q. We write*

$$P \subset Q \tag{4.28}$$

If P has at least one element which is not in Q then P is not a subset of Q and we write

$$P \not\subset Q \tag{4.29}$$

Equation 4.26 means that P is either a proper subset of Q or is equal to Q. Equation 4.28 means that Q is a ***superset*** *of P and write $Q \supset P$.*

This definition allows for P or Q or both to be the empty set. That possibility brings some complications of interpretation, which we will consider shortly. It also allows for P and/or Q to be infinite.

For now, assume that both P and Q are non-empty.

The notation $\supset$ for **superset** is just subset $\subset$ backwards.

The relations of definition 4.7 are illustrated in the figure below.

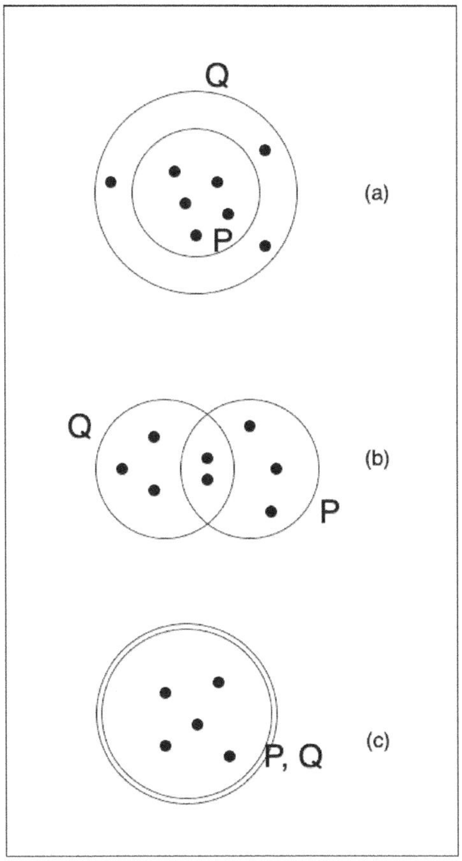

Figure 4.4:

In this figure: (a): $P \subset Q$, (b): $P \not\subset Q$, (c): $P = Q$.

The following uses the sets equation 4.1. Start with proper subset, equation 4.28. Consider

$$A \subset F \qquad (A = \{1, 3, 4, 7, 10\}, \quad F = \{1, 3, 4, 7, 10, 12, 15\}) \tag{4.30}$$

In this example, every element of A is in F. However, F has at least one element which is not in A (in this case, the two elements 12, 15 of F are not in A). Therefore A is a **proper subset** of F. F is a proper **superset** of A, $F \supset A$.

Note that, for **any** two sets P, Q, if $P \subset Q$, then by definition Q cannot be empty (because, by the definition above, Q must have at least one element not in P). Set P could be the empty set.

Now consider equality, equation 4.27.

$$A = G \qquad (A = \{1, 3, 4, 7, 10\}, \quad G = \{1, 3, 4, 7, 10\}) \tag{4.31}$$

Since sets A and G have exactly the same elements, they are equal.

Now consider subset (not necessarily proper), equation 4.26. Here, **we must follow the *rule* of definition 4.7 exactly.** Referring to equation 4.31, every element of set A is in set G and therefore, strictly speaking, A is a subset of G. However, in this example, A is not a proper subset of G because G does not have an element which is not in A. Therefore A is a subset of G and also is equal to G (as in the above example), so $A \subseteq G$.

In formulating and proving theorems, equation 4.26 takes into account the **possibility** that a set P may be a proper subset of Q or may be equal to Q. For example, we might state a theorem by postulating two sets P, Q such that $P \subseteq Q$. This means that the theorem is applicable to a set P which is either a proper subset of Q **or** is equal to Q. If on the other hand, if our theorem requires that P be a proper subset of Q, then we would specify $P \subset Q$.

From eqs 4.1

$$A \subset F, \quad A = G, \quad A \subseteq G, \quad A \not\subset B, \quad B \not\subset F, \quad A \not\subset G \tag{4.32}$$

Exercise: Verify equation 4.32.

Theorem 4.8

Given two sets P, Q, with $P \subset Q$ (so Q not empty). Then

$$P \cap Q = P, \quad P \cup Q = Q \tag{4.33}$$

This theorem is valid even if $P = \emptyset$. Exercise: Prove this theorem (use a diagram).

Let us now consider the situation of definition 4.7 when $P = \emptyset$, $Q \neq \emptyset$. That is, P is empty and Q is any set which is not empty. Considering the rule for equation 4.28, we see that the condition for $P \subset Q$ is that every element of P is in Q. The *standard* argument is as follows. Since P has no elements, then all its elements are indeed in Q. Therefore, **the empty set is a proper subset of any non-empty set Q.**

$$\emptyset \subset Q, \quad Q \text{ not empty} \tag{4.34}$$

I will concede that reasoning, which is found in almost all non-formal treatments of subsets, is not compelling. Now suppose that both P and Q are empty. Since $\emptyset = \emptyset$, the same argument as above indicates that $\emptyset \subseteq \emptyset$. So we have established the following theorem.

Theorem 4.9

For any non-empty set Q

$$\emptyset \subset Q \tag{4.35}$$

In addition

$$\emptyset \subseteq \emptyset \tag{4.36}$$

The empty set is a subset of any set.

There are some theorems in set theory which may seem peculiar, or possibly even untrue, but which are actually **edge conditions** of a definition or theorem. Such edge conditions are useful because, in many cases, they eliminate the necessity of including a lot of special cases in theorems or definitions. Theorem 4.9 can be said to be one such theorem.

Consider the following.

$$P \cap Q \subseteq P, \quad P \cap Q \subseteq Q \tag{4.37}$$

This is obvious if P and Q are non-empty, since everything in $P \cap Q$ is in P and also in Q. If P or Q is empty, then theorem 4.9 is seen to be an *edge condition* of equation 4.37

It is certainly unclear how to intuitively interpret equation 4.35. What exactly does it mean that a set with nothing is a subset of a set with something? It probably means that this author should develop a more liberal concept of empty set, although he finds it difficult.

Theorem 4.9 is integral to applications of informal set theory. Later we present additional arguments to justify this result, and show that theorem 4.9 is required for consistency, regardless of whether it intuitively means something or not. The following is another edge condition.

Theorem 4.10

Any set is a subset of itself.

In other words, for any sets A, B, C

$$A \subseteq A, \quad B \subseteq B, \quad C \subseteq C, ... \tag{4.38}$$

To show this, we just follow the rule expressed in equation 4.26. Is every element of A (on left) in A (on right)? Yes. Is every element of A (on right) in A (on left). Yes. Therefore $A \subseteq A$. Similarly for B, C, and so forth.

On the other hand, it is **not true** that $A \subset A$. **A is not a proper subset of itself**, because, following the rule of equation 4.28, there is no element of A (on the right) which is not in A (on the left).

Note that the following two expressions are *inconsistent*:

$$P \subset Q \qquad Q \subseteq P \quad (inconsistent) \tag{4.39}$$

Since P is a proper subset of Q, then Q must have an element which is not in P. Therefore Q cannot be a subset of P. However, we have the following important theorem, which is used in many proofs.

Theorem 4.11

P, Q any sets. If $P \subseteq Q$ and $Q \subseteq P$ then $P = Q$.

To prove this: If Q has an element not in P then $Q \not\subset P$, a contradiction of an assumption. Similarly the reverse. Therefore P and Q have the same elements.

Note that **no non-empty set is a subset of the empty set.** For A, B not empty

$$A \not\subset \emptyset, \quad B \not\subset \emptyset, ... \tag{4.40}$$

We follow the rule of equation 4.29. The set A has at least one element which is not in the empty set, and so does B (and so does any other non-empty set).

5 defining more general sets

5.1 defining sets of integers by using sequences

It would be inconvenient (to say the least) to specify a set containing all integers between 1 and 7,000 by writing all 7,000 numbers within curly brackets. Here is a shortcut .

$$A = \{1, 2, 3, ..., 7000\} \tag{5.1}$$

This device is used to create a set whose elements conform to a pattern. Between brackets, start with the first few numbers of the sequence, which sets the pattern, add a few dots, then enter the final number of the sequence. Here is a set with even integers between 1 and 5000

$$B = \{2, 4, 6, ..., 5000\} \tag{5.2}$$

We have:

$$A \cup B = A, \quad A \cap B = B, \quad B \subset A, \quad A \not\subset B \tag{5.3}$$

As another example, the set of integers divisible by 3 between 21 and 333 is the set $\{21, 24, 27, ..., 333\}$.

5.2 specifying sets algorithmically

We have already seen how to create a set from two or more other sets either by union or intersection. Here we show how to create a new set based on only one other set.

Consider the set

$$A = \{1, 2, 3, ..., 100\} \tag{5.4}$$

and suppose we want to create a set of the squares of these integers, where we only want squares between 1 and 900 inclusive(the square is less than or equal to 900). We could use brute force and write the set as $\{1, 4, 9, 16, 25, ..., 900\}$, but in more complicated cases the pattern may not be obvious. For example, what about a set of squares between 297^2 and 1342^2. There is an easier way, which generalizes nicely, and which, in the end, is a nice way to demonstrate theorem 4.9.

The idea is to **conceptually** create a set one element at a time - element by element - according to an algorithm. The set C of the squares of positive integers between 1 and 900 may be created from the set A, equation 5.4, as follows:

$$C = \{x^2 | \forall x \in A \ and \ x^2 \leq 900\} \tag{5.5}$$

The symbol $\forall$ means **for all** or **for each**. The vertical line means **such that** or **where** or **when**. The element to be inserted into set C appears to the left of the vertical line.

The above expression is read as follows: Start with $C = \emptyset$, the empty set. The expression $\forall x \in A$, when it appears in a set definition, means that every positive integer in the set A (in this case equation 5.4), one at a time, is tested. The test in this case is whether the square of that integer is less than or equal to 900. The x^2 in front of the vertical bar means x^2 is inserted into C. Taken all together, for each $x \in A$, insert x^2 into C only when $x^2 \leq 900$.

The set C is *constructed* element by element. The steps are:

0. Let $C = \emptyset$.
1. Get a new positive integer x from A.
2. If x^2 is less than or equal to 900, insert x^2 into C
3. If not all integers in A are tested yet, go to step 1.
4. Stop.

Step 0 is absolutely crucial. At step 1, a **new** positive integer is obtained from A. Step 2 is a test-and-insert, where insertion into a set is defined in definition 4.4. Step 3 tests whether all the elements of set A have been tested. Step 4 is the end.

Here is a simpler example. For set A, equation 5.4

$$D = \{x|\forall x \in A \, and \, x \geq 10 \, and \, x < 50\} = \{10, 11, 12, ..., 49\} \tag{5.6}$$

This creates a proper subset $D \subset A$.

The notation can be streamlined. The notation here is borrowed from mathematical logic, and an introduction to logical notation is given in the section *symbolic logic and set theory.* (Here we just use a subset of the concepts, sufficient for now.) Considering equation 5.5, sometimes the $\forall$ symbol is omitted, so that the expression is written simply as

$$C = \{x^2|x \in A \, and \, x^2 \leq 900\} \tag{5.7}$$

Instead of writing *and* one uses the symbol & so that

$$C = \{x^2|x \in A \; \& \; x^2 \leq 900\} \tag{5.8}$$

However, one **cannot** have

$$C = \{x^2|x^2 \leq 900\} \qquad (incorrect) \tag{5.9}$$

The problem with this is that we do not know what kind of element x is. It could be an integer, a spaceship, a planet, a real number - anything at all. This kind of construction may lead to paradox, as in the section *Russell's paradox.* For now, the following is most important **in practice** (that is, when using set theory)..

> When creating sets using the constructions of this section, $\{something \mid x \in X \; \& \; ......\}$, the set X must be well defined and exist. In addition, the test (represented by the dots ....) must be valid for every element of X.

Equation 5.6 in this streamlined notation is

$$D = \{x|x \in A \;\&\; x \geq 10 \;\&\; x < 50\} = \{10, 11, 12, ..., 49\} \tag{5.10}$$

Here is another example, using set A, equation 5.4 ($A = \{1, 2, 3, ..., 100\}$).

$$S = \{x|x \in A \;\&\; x^2 < 90\} \tag{5.11}$$

The result is $S = \{1, 2, 3, 4, 5, 6, 7, 8, 9\}$. Note that $S \subset A$.

We can create a set of even integers as follows. The notation $a|b$ is a **logical statement** which is true if 'a' divides 'b' and false otherwise. For example, $2|12$ is true, $3|14$ is false, $7|21$ is true, $16|21$ is false. Then

$$V = \{x|x \in A \;\&\; 2|x\} \tag{5.12}$$

The result is $V = \{2, 4, 6, ...100\}$. The only values of x that are in V are the ones where $2|x$ is true (ie, where x is even). Note $V \subset A$.

We also have

$$A = \{x|x \in A\} \tag{5.13}$$

Set creation similar to equations 5.10, 5.11, 5.12, 5.13 are important because their generalization leads to a justification of theorem 4.9.

Theorem 5.1

Let P be any set, including the empty set. Define a new set from the elements of P as follows.

$$aSet = \{x|x \in P \;\&\; some\, conditions\, on\, x\} \tag{5.14}$$

then

$$aSet \subseteq P \tag{5.15}$$

The reason $aSet$ is a subset of P is that all the elements added to $aSet$ are from P. If, for some x, the condition on x is false, that element will not be in $aSet$.

This seems simple enough. So let's consider the following legitimate (though peculiar) situation.

$$E = \{x|x \in A \;\&\; x^2 < 0\} \tag{5.16}$$

where A is defined by equation 5.4. The procedure for creating this set is:

- $E = \emptyset$
- Get a new positive integer x from A
- If $x^2 < 0$ insert x into E
- If not all integers are tested, go to step 1
- Stop

The problem of course is that, based on step 3, **no elements are inserted into E**, so that at the end, E is still the empty set. But theorem 5.1 states that $E \subseteq A$. Since A is non-empty and $E = \emptyset$, then $\emptyset \subset A$.

We can make the same argument for **any** set P of theorem 5.1. If the *conditions on x* are false for all $x \in A$, then $aSet$ must be the empty set and $aSet \subset P$.

From this, theorem 4.9 follows.

This would seem to be a reasonable justification for theorem 4.9. **It does indicate that that theorem is more a consistency requirement than a fundamental theorem about the nature of sets.** It is still difficult for this author to develop an intuition about an empty set as a subset.

Some authors believe that the empty set is a convenience and that all of mathematics can be developed from set theory without using the empty set. I don't know whether or not this is true, but it seems that theorems would require many extra conditions on whether sets are empty or not.

Here is an example of set creation illustrating the no-repeat condition on sets.

Let $B = \{-1, 1, -2, 2, -3, 3\}$. The set

$$H = \{x^2 | \forall x \in B \, and \, x^2 < 900\} = \{1, 4, 9\} \tag{5.17}$$

In this case, the algorithm would seem to try to insert 1 into H twice, since $(-1)^2 = (1)^2 = 1$. The same with 4 and 9. However, repeats are not allowed in the set H, so **implicit** in this procedure is that the resultant set has no repeats. A given element can only be inserted once.

5.2.1 the set-creating-process as computer program

The process illustrated above is very much like a computer program, which would look something like this (for a generic object oriented language): Define some set Z of integers and invoke the function.

$$C = createSetOfSquares(SetZ, int\, maximumSquare) \tag{5.18}$$

with the definition of the function **createSetOfSquares**

```
Set createSetOfSquares(Set Z, int maximumSquare)
{
   Set newSet;  /*creates newSet = empty set */
   int x;
   forEach x in Z
      if(x^2 <=maximumSquare)newSet.insert(x^2);
   return newSet;
}
```

Note that, if Z does not contain any integers whose square is less than or equal to 900, the program will return an empty set. The **insert** *method* for a Set object is presumed to be such that the same element cannot be added to a set more than once. (No repeats.)

5.3 sets with indexed elements

We can create sets some (or all) of whose elements have **subscripts** (called **indices**).Consider the set

$$A = \{x_1, x_2, x_3, x_4\} \tag{5.19}$$

Here each x_i may be either an abstract element or a variable. Because there are no repeats, we must have $x_i \neq x_j$ for all i, j, when $i \neq j$. Generally all elements are of the same type, although we certainly could mix and match. Here is a toy example.

$$\begin{aligned} &a_1 = 3,\;\; x_3 = 7 \\ &S = \{1, 2, x_1, x_2, x_3, x_4, a_1, a_2, a_3\} \end{aligned} \tag{5.20}$$

The result is $S = \{1, 2, 3, 7, x_1, x_2, x_4, a_2, a_3\}$ where the indexed elements have not been assigned and therefore are abstract elements.

If we know that we have a set of indexed elements $w_1, w_2, w_3, ..., w_N$, all distinct, then we can define a set F based on a sequence of the **indices**,

$$F = \{w_1, w_2, w_3, ..., w_N\} \tag{5.21}$$

Of course the indices don't have to be in any order or start from 1. Here is a legitimate set

$$S = \{w_{12}, w_{99}, w_2, w_7\} \tag{5.22}$$

5.3.1 sets whose names are indexed

The names of sets can have subscripts. Consider the example

$$\begin{aligned} A_1 = \{1,2,3\},\ A_2 = \{4,a,b\},\ A_3 = \{x_1,x_2,x_3\} \\ B = A_1 \cup A_2 \cup A_3 = \{1,2,3,4,a,b,x_1,x_2,x_3\} \end{aligned} \tag{5.23}$$

5.4 subscripts can be anything

Suppose we have some abstract entities a, b, c, d,.... which may be in some set. Then it is legitimate to define variables or other abstract entities as follolws:

$$S = \{w_a, x_b, y_c, w_d\} \tag{5.24}$$

where each element is distinct. If a, b, c, d are variables, this is a kind of *indirect indexing.*

The same goes for subscripts on sets. Reusing equation 5.23 we could have

$$\begin{aligned} A_p = \{1,2,3\},\ A_q = \{4,a,b\},\ A_r = \{x_1,x_2,x_3\} \\ B = A_p \cup A_q \cup A_r = \{1,2,3,4,a,b,x_1,x_2,x_3\} \end{aligned} \tag{5.25}$$

Here, p, q, r may be variables or abstract entities - anything so long as they are distinct.

5.5 index sets

Consider the example of the previous section, equation 5.21

$$F = \{w_1, w_2, w_3, ..., w_{20}\} \tag{5.26}$$

We may define an **index set I**

$$I = \{2, 8, 14, 19\} \tag{5.27}$$

Using this set I, we may create a set whose elements w_k are in F and *also* such that k is in I, the index set ($k \in I$). (The w_k must be defined somewhere.) So

$$S = \{w_k | w_k \in F \,\&\, k \in I\} = \{w_2, w_8, w_{14}, w_{19}\} \tag{5.28}$$

This works as follows. Start with S being the empty set. Loop through all elements w_k of F, and if $k \in I$, then add w_k to S. Of course if I were the set $I = \{100, 200, 300\}$, then S would be the empty set. This obviously generalizes

Definition 5.2

Let $F = \{w_1, w_2, ..., w_N\}$ where w_i are all distinct. Let I be ***any*** *set of integers but* ***excluding*** *the empty set. Then we may create a set G from elements of F, the elements being chosen based on the index set I.*

$$G = \{w_k\}_{k \in I} = \{w_k | w_k \in F \,\&\, k \in I\} \subseteq F \tag{5.29}$$

The first expression is a ***shorthand way of writing a set whose elements have subscripts****.*

The case of I being the empty set is tricky, and is discussed in the section on symbolic logic.

There can be problems. Suppose we have defined abstract elements $w_1, w_2, ..., w_{20}$ and an index set $I = \{2, 5, 12, 18, 34, 980\}$. If we try to define a set $G = \{w_2, w_5, w_{12}, w_{18}\}$ with the notation

$$G = \{w_k\}_{k \in I} \tag{5.30}$$

there is an obvious problem, in that w_{34} and w_{980} are not defined. Obviously this is an error. If the user is careful, such a situation should not occur. Or the user could specify that if w_k is not defined for some k, then ignore it. Or else some other procedure.

digression - it's a bug. This situation is analogous to array overflow in a computer program, when an array is addressed indirectly.
end digression

The best approach, obviously, is to make sure everything is defined.

An index set need not contain numbers. So $I = \{a, b, c\}$ is an index set, and if $A = \{w_a, w_b, w_c, w_d, w_e, w_f, \}$, then

$$\{w_\rho | w_\rho \in A \;\&\; \rho \in I\} = \{w_a, w_b, w_c\} \tag{5.31}$$

5.6 set of sets. subsets

It is very common that the some or even all of the elements of one set are themselves sets. Define the set S as

$$S = \{\{1\}, \{1, 2\}, \{2, 3\}, 1, 3, 4\} \tag{5.32}$$

Some of the elements of S are sets, and some are not sets. The sets $\{1\}, \{1, 2\}, \{2, 3\}$ are **elements** of S, as are the numbers 1 and 3 and 4. (Note that $\{1\} \neq 1$, so elements of S are distinct.) Therefore

$$\begin{aligned} &\{1\} \in S, \quad \{1, 2\} \in S, \quad \{2, 3\} \in S \quad 1 \in S, \quad 3 \in S, \quad 4 \in S \\ &2 \notin S, \quad \{1, 3\} \notin S, \quad \{1, 2, 3\} \notin S \end{aligned} \tag{5.33}$$

A set all of whose elements are sets is sometimes called a **collection**, so it would be a *collection of sets*

What are the subsets of S? A subset of S is a set containing some or all of the **elements** of S (the empty set is also a subset). In this case, some of the elements of S are themselves sets, so a subset of S could also contain sets.

$$\{\{1\}\} \subset S, \quad \{\{1,2\}\} \subset S,, \quad \{\{1\}, \{2,3\}\} \subset S, \quad \{1, 4, \{1,2\}\} \subset S \tag{5.34}$$

and so forth. Note all the extra parentheses. For example, $\{1\}$ is an **element** of S, and therefore that element (a set) would be a *member* of a subset $\{\{1\}\} \subset S$. Note

$$1 \not\subset S, \quad 1 \in S, \quad \{1\} \in S, \quad \{1\} \subset S, \quad \{\{1\}\} \subset S, \quad \{\{1\}\} \notin S \tag{5.35}$$

Note that we have already discussed $\emptyset \subset S$. However

$$\emptyset \notin S \tag{5.36}$$

That is, S does not contain the empty set explicitly.

However, a set may indeed contain the empty set. Let

$$A = \{0, \emptyset, \{0\}\} \tag{5.37}$$

Then

$$\begin{aligned} &\emptyset \in A, \quad \emptyset \subset A, \quad 0 \not\subset A, \quad 0 \in A, \quad \{0\} \in A \\ &\{0\} \subset A, \quad \{\{0\}\} \subset A, \quad \{0, \emptyset\} \subset A, \quad \{0, \{0\}\} \subset A \end{aligned} \tag{5.38}$$

It is very important to understand the expressions above.

As another example, consider the sets

$$\begin{aligned} &A = \{a, b, c, S\}, \quad S = \{a, b, c\} \\ &B = \{a, b, S\} \end{aligned} \tag{5.39}$$

Then $S \subset A$ because $S = \{a, b, c\} \subset \{a, b, c, S\}$. However, $S \not\subset B$ because $c \notin B$. The fact that S is a subset of A is not because $S \in A$ but because all the elements of S are members of A individually. Also, $\{\{a, b, c\}\} = \{S\} \subset A$ and $\{S\} \subset B$. In the latter case, the set $S = \{a, b, c\}$ is an **element** of B and thus $\{S\}$ is a subset of B. These constructions are important for understanding transitive sets, which are introduced in the section on ordinal numbers.

We can also have a collection whose elements are sets with indices. Suppose sets $A_1, A_2, A_3, ..., A_{10}$ have been defined. Then we can define collections consisting of all these sets or only some of them.

$$\begin{aligned} &S = \{A_1, A_2, A_3, ..., A_{10}\} \\ &I = \{2, 3, 6\} \\ &X = \{A_i\}_{i \in I} = \{A_1, A_2, A_6\} = \{A_i | A_i \in S \;\&\; i \in I\} \\ &X \subset S \end{aligned} \tag{5.40}$$

5.7 unions and intersections of sets of sets - notation

This section describes notations for union and intersections for sets of sets. This section can be skimmed on a first reading.

Define

$$A_1 = \{1,3,5,8\}, \quad A_2 = \{a,b,c\}, \quad A_3 = \{x_1, x_2\} \tag{5.41}$$

The union of these sets can be written

$$R = A_1 \cup A_2 \cup A_3 = \bigcup_{i=1}^{3} A_i = \{1,3,5,8,a,b,c,x_1,x_2\} \tag{5.42}$$

The union follows the algorithm of section 4.1.2 (based on definition 4.2). Start off with $R = \emptyset$. Then as i iterates from 1 to 3, add the elements of set A_i to R, excluding repeats.

> When there is an iteration such as
>
> $$\bigcup_{i=a}^{b} \tag{5.43}$$
>
> with $a < b$, integers, this implies that the iteration variable takes on integer values. However, these symbols may also be used for non-integers, so context will be important.

Here is another notation. Let $S = \{A_1, A_2, A_3\}$. Then we also can write

$$R = A_1 \cup A_2 \cup A_3 = \bigcup_{X \in S} X = \{1,3,5,8,a,b,c,x_1,x_2\} \tag{5.44}$$

In this notation, X is an iteration variable and takes on the values A_1, A_2, A_3 in succession, and elements of each A_i are added to R one after another.

This notation holds for any kind of set, even infinite ones, and even those where the elements of the set do not have subscripts. For example, suppose $T = \{P, Q, R, W\}$ a collection of four sets. Then $R = \bigcup_{X \in T} X$ is a set R with all the elements of P and Q and R and W.

Finally we can also write (set S given above equation 5.44)

$$R = A_1 \cup A_2 \cup A_3 = \bigcup S = \{1,3,5,8,a,b,c,x_1,x_2\} \tag{5.45}$$

where the unindexed $\bigcup S$ means to take the union of all sets of S. This notation is valid only when all elements of S are sets.

Definition 5.3

Let $S = \{A_1, A_2, ..., A_N\}$, each A_i a set. There are four notations for the union of all A_i.

$$A_1 \cup A_2 \cup .... \cup A_N = \bigcup_{i=1}^{N} A_i = \bigcup_{X \in S} X = \bigcup S \tag{5.46}$$

It can be tricky. Define $P = \{1, 2\}$, $Q = \{ \{a, b\}, \{c, d\} \}$, $S = \{P, Q\}$. Then

$$\bigcup S = \{1, 2, \{a, b\} \{c, d\} \} \tag{5.47}$$

The notation can be generalized to incorporate index sets. Let $S = \{A_1, A_2, ..., A_{10}\}$, A_i sets, and define the index set $H = \{1, 3, 7\}$. Then

$$A_1 \cup A_3 \cup A_7 = \bigcup_{k \in H} A_k \neq \bigcup S \tag{5.48}$$

One can use subscripts which are not integers. Let $T = \{A_a, A_b, A_c, ..., A_z\}$ and define the index set $J = \{a, c, f, m\}$. Then

$$A_a \cup A_c \cup A_f \cup A_m = \bigcup_{\rho \in J} A_\rho \tag{5.49}$$

There is another notation which will be useful. Consider

$$\{1, 2, 3, ..., 10\} = \bigcup_{k=1}^{10} \{k\} \tag{5.50}$$

This is exactly the same as if we were to define $A_k = \{k\}$ and then

$$\{1, 2, 3, ..., 10\} = \bigcup_{k=1}^{10} \{k\} = \bigcup_{k=1}^{10} A_k \tag{5.51}$$

The same notation of this section is also used for intersections, with $\cap$ replacing $\cup$.

6 infinite sets - an introduction

6.1 what is infinity?

So far we have discussed finite sets. A finite set is a set with a finite number of elements. An infinite set is defined as follows:

Definition 6.1

An ***infinite set*** *is a set which is not finite.*

The set of all positive integers is referred to as $\mathbb{Z}_+$. One way to notate it is as follows. Start off with the sequence, as in section 5.1, but leave off the end. The set

$$\mathbb{Z}_+ = \{1, 2, 3, 4, ...\} \tag{6.1}$$

denotes the set of all positive integers. The set $\{2, 4, 6, 8, ...\}$ denotes the set of all even integers.

Theorem 6.2

The set of positive integers $\mathbb{Z}_+$ is infinite (that is, not finite).

Suppose Z_+ were finite. Then it would have a maximum element. That is, some integer in the set would be the largest integer in the set. Say this integer is N. But since $\mathbb{Z}_+$ contains **all** integers, then N+1 must also be in the set, a contradiction (since N was supposed to be the largest element of the set). Therefore, the set of integers is not finite, which, by definition, means it is infinite.

A similar argument holds for the set of even integers.

An infinite set of abstract elements may be notated as

$$\{w_1, w_2, w_3, ...\} \tag{6.2}$$

where the sequence is made up of the indices and all w_i are distinct.

How should we think about an infinite set? In the case of $\mathbb{Z}_+$, we can say the following: If n is any positive integer, then $n \in \mathbb{Z}_+$. In other words, every **finite** integer is in the set. But, by definition, **all integers are finite**. Therefore, $\mathbb{Z}_+$ **is an infinite set where its every element is finite**.

It can get more complicated. Suppose we remove the number 8 from $\mathbb{Z}_+$. The resulting set is obviously still infinite, but it is no longer the set of all positive integers. Therefore we can no longer state that for any integer n, $n \in \mathbb{Z}_+$. However, we can still say that every integer in this set is finite.

What about the infinite set $W = \{w_1, w_2, w_3, ...\}$? The w_i are abstract elements, so it is meaningless to say that every element of W is finite. What we would say is that every element in W is associated with a **finite subscript** i. There is no such element as w_∞.

What about the symbol ∞? Actually ∞ **is only a symbol**. It is not a number in the same way that a finite number is understood. We could consider it similar to an abstract element, although it stands for *something* which may be intuitive but which is not well defined. Since $\mathbb{Z}_+$ is a set of *numbers*, and ∞ is a symbol (and not a number), then ∞ cannot be an element of $\mathbb{Z}_+$. That is

$$\infty \notin \mathbb{Z}_+ \tag{6.3}$$

This equation is a metaphor only unless ∞ is considered an abstract object. Consider the following infinite set

$$A = \{1, \frac{1}{2}, \frac{1}{3}, \frac{1}{4}, ...\} \tag{6.4}$$

which can be defined as

$$A = \{\frac{1}{n} | n \in \mathbb{Z}_+\} \tag{6.5}$$

Based on this definition, we see that, if n is **any** positive integer, then $\frac{1}{n} \in A$. So it is reasonable to conclude that A is an infinite set. Note that

$$0 \notin A \tag{6.6}$$

because there is no integer n such that $\frac{1}{n} = 0$ (all integers are finite). The expression $\frac{1}{\infty} = 0$ is meaningless. Therefore

$$\{0\} \cap A = \emptyset \tag{6.7}$$

We should discuss the 'creation' of the set of equation 6.5 using the algorithm of section 5.2. Based on our discussion of algorithmic set creation, equation 6.5 requires an infinite loop through the set $\mathbb{Z}_+$, which certainly is not possible in real time. No matter how fast one imagines going through the loop, for example, it can't be done in a finite time.

One must somehow come to terms with this procedure imaginatively. We could simply state that this set exists and leave it at that (which may be the best thing). But it would be nice to have some intuition about what goes on in the imagination. One idea, which is sort of science-fictiony, or at least poetic, is to simply state that the imagination can compress an infinite amount of time to an instant, so the infinite loop can be done instantaneously. It's not clear that the problem is well-defined in the context of this book.

One can consider an infinite iteration in terms of verification. Regarding equation 6.5, is $\frac{1}{20}$ in A? Yes. Is 22 in A? No.

Another aspect of an infinite set is the things it contains. Consider some very large numbers, as in section 3.16. **The set $\mathbb{Z}_+$ contains very large integers.** There are integers with a billion **digits**, a trillion digits, and integers with a trillion trillion digits and integers which make these integers look tiny, and then even larger ones.

We may form the set

$$B = \{1, 2, 3, ..., 378199203898675364781293 7 \times 10^{150}\} \tag{6.8}$$

One might remark that this is a very large set indeed. However, it is a finite and not an infinite set, because there exists an integer n not in B

$$n = 1 + 3781992038986753647812937 \times 10^{150} \notin B \tag{6.9}$$

On the other hand, for set A defined in equation 6.5

$$\frac{1}{1 + 3781992038986753647812937 \times 10^{150}} \in A \tag{6.10}$$

because the fraction, while conventionally considered very very very small, is in fact not zero.

Such large numbers are inconceivable except in the abstract. What would be their purpose? When would they be used? Their existence is justified as necessary in order to create infinite sets of numbers and then prove theorems about them and involving them (for example, fractions such as equation 6.10 required for the epsilon-delta analysis of the function $\frac{\sin x}{x}$ for small |x|).

6.2 infinite union of integers

We may consider $\mathbb{Z}_+$ as an index set and define the set of even integers EV as

$$EV = \{2k\}_{k \in \mathbb{Z}_+} = \bigcup_{k \in \mathbb{Z}_+} \{2k\} \tag{6.11}$$

(see equation 5.50 for notation). EV is an infinite set, and if k is in $\mathbb{Z}_+$ then 2k is in EV. There is an algorithmic definition

$$EV = \{2x | x \in \mathbb{Z}_+\} = \{x | x \in \mathbb{Z}_+ \,\&\, 2|x\} \tag{6.12}$$

Sometimes one will see EV defined as

$$EV = \bigcup_{k=1}^{\infty} \{2k\} \tag{6.13}$$

Here ∞ is not a number but a symbol, and obviously the **non-number** 2∞ is not added to EV, a set of numbers. What the symbol ∞ means in a context such as equation 6.13, is that, for every integer k, EV contains the integer 2k. The upper limit of ∞ means k loops through all integers, but is never set equal to ∞ (that would be meaningless). Equation 6.13 is exactly equivalent to equation 6.11.

6.3 comparing infinite sets

Consider the sets

$$\begin{aligned} \mathbb{Z}_+ &= \{1,2,3,4,5,6,7,8.....\} \\ H &= \{2,4,6,8,10,12,14,16,...\} \\ J &= \{5,10,15,20,25,30,35,40,....\} \end{aligned} \tag{6.14}$$

The relationships between these sets are illustrated in the following figure.

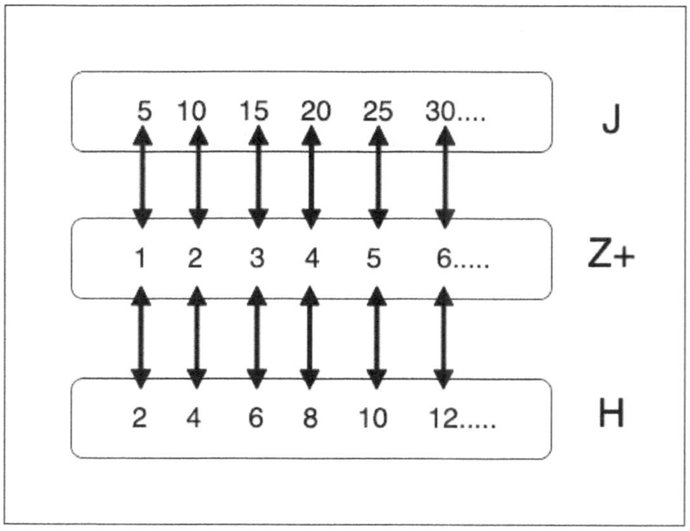

Figure 6.1:

We see that, for example, the sets $\mathbb{Z}_+$ and H are in one-to-one correspondence $1 \leftrightarrow 2, 2 \leftrightarrow 4, 3 \leftrightarrow 6$ and so forth. Both sets are infinite, so it appears that there are as many even numbers as there are integers. Since we naively expect that there should be fewer even numbers than there are integers, we seem to have a **paradox**. Galileo made the same curious observation.

In fact the paradox is even stranger because

$$H \subset \mathbb{Z}_+ \tag{6.15}$$

so H a **proper** subset of $\mathbb{Z}_+$. So now we have a proper subset being in 1-to-1 correspondence with its superset.

Such a thing cannot happen for finite sets. Consider

$$A = \{1,2,3,...,10\}, \quad B = \{2,4,6,...,20\} \tag{6.16}$$

These sets are in 1-1 correspondence $(1 \leftrightarrow 2, 2 \leftrightarrow 4, ...)$.However

$$B \not\subset A \tag{6.17}$$

which is different from equation 6.15. Infinite sets behave differently than finite sets.

Similarly the elements of H and J, and of $\mathbb{Z}_+$ and J, are in one-to-one correspondence. There are as many integers divisible by five as there are even integers or integers, and yet J is a proper subset of $\mathbb{Z}_+$.

Still, it is the case that $\mathbb{Z}_+ \neq H$, $\mathbb{Z}_+ \neq J$, $H \neq J$ and so forth because, for example, there is an element in $\mathbb{Z}_+$ which is not in J and not in H.

We *resolve the paradox* by appreciating that we are (inappropriately) thinking of these infinite sets in the same terms that we use for finite sets. Clearly, infinite sets behave very differently than finite sets. While it is true that the definitions of union, intersection, subset are valid for infinite sets, nonetheless the **cardinality** (ie, size) of an infinite set cannot be computed by counting the elements. There is only one symbol, ∞, for infinity, and we have demonstrated already three unequal infinite sets H, J, $\mathbb{Z}_+$ which are in 1-1 correspondence. So saying that $\mathbb{Z}_+$ is **larger** than the set of even integers does not seem to be entirely meaningful.

We need a new definition of something which will be called the *cardinality* of an infinite set, which bears some relationship to size, but which does not count elements. And we need some quantitative way of distinguishing finite from infinite sets and some way comparing two or more infinite sets besides drawing pictures such as figure 6.1.

In order to understand this behavior, and to be able to make sense of infinite sets in general, we have to know something about **maps** (functions), which is the subject of the next section.

7 maps (AKA functions)

Maps are the crucial component in allowing a quantitative analysis of infinite sets.

A single set is not particularly interesting by itself. Rather, what turns out to be very interesting, and very useful, is the comparison of one set with another, or a transformation from one set to another. The comparisons or transformations are accomplished by means of **maps**, also called **functions**. We have seen in the previous section that, while the set of all integers is mildly interesting, what is really interesting is the paradox resulting from the comparison of the set of integers with the set of even integers. This is where functions come in.

In set theory, maps are defined abstractly, or formally, in terms of **ordered pairs**. We will introduce that definition, and also use the more common function notation used in analysis. The introduction of ordered pairs will not be wasted as they appear in many places in set theory.

7.1 ordered pairs

Let a, w be two elements of the same or different sets. An **ordered pair** of a and w is notated (a,w). This is also called a **2-tuple**. An ordered pair has no intrinsic interpretation. Its interpretation resides in the way that it is used.

For the ordered pair (a, w), the element a is in the **first slot** of the ordered pair, the element w in the **second slot**.

If a, w are distinct (that is, $a \neq w$), then (a, w) is **not** the same as (w, a). The ordered pair (a, a) is perfectly valid, so *the two elements of an ordered pair can be the same.*

Two ordered pairs, (a,w) and (b,x) are equal - $(a, w) = (b, x)$ - if and only if if
a = b and w = x. Otherwise they are not equal.

Ordered pairs may be elements of a set. For example, here is a set of ordered pairs

$$S = \{(1, a), (2, q), (5, w), (2, x)\}, \quad x \neq q \tag{7.1}$$

A set such as S is frequently referred to as a **collection**, in this case a collection of ordered pairs. Each ordered pair is a single element of the set. Each ordered pair in a collection must be distinct. In set S above, $x \neq q$ so (2,q) and (2,x) are distinct ordered pairs.

Sets of ordered pairs behave just like sets of anything. There is set inclusion

$$(1, a) \in S, \quad (5, w) \in S \tag{7.2}$$

While an ordered pair is an element of S, the contents of the individual slots are not.

$$a \notin S, \quad 5 \notin S \tag{7.3}$$

The union of collections of ordered pairs is an ordinary union of elements.

$$\{(a,q),(c,w),(1,4)\} \cup \{(r,5),(a,q)\} = \{(a,q),(c,w),(1,4),(r,5)\} \tag{7.4}$$

with no repeat of (a,q). We can also have subsets of ordered pairs:

$$\{(5,w),(2,x)\} \subset S \tag{7.5}$$

Let's not forget that $\emptyset \subset S$. However, $(\emptyset, \emptyset) \not\subset S$.

A set of ordered pairs can be thought as two columns of a spreadsheet or of a table, one column for values in the first slots, the second column for values in the second slots.

7.1.1 an ordered pair may be defined as a set

We may define

$$(a,b) = \{a,\{a,b\}\} = \{a,\{b,a\}\} \tag{7.6}$$

This has the property that, if $a \neq b$,

$$(a,b) = \{a,\{a,b\}\} \neq \{b,\{a,b\}\} = (b,a) \tag{7.7}$$

7.2 map defined

Maps are defined in terms of ordered pairs and also in terms of the usual function notation.

Definition 7.1 *Let F be a* ***map*** *from set P to set Q, with both P, Q not empty. We write*

$$F : P \to Q \tag{7.8}$$

The map F is a collection (a set) of ordered pairs (section 7.1).

$$F = \{(p_1,q_1),(p_2,q_2),....\}, \quad p_i \in P, q_i \in Q \tag{7.9}$$

such that:

1. *Every element of P must be in the first slot of one and only one ordered pair in F. Therefore each ordered pair is distinct.*

2. *Not every element of Q need be in the second slot of the ordered pair.*

An alternative notation (strictly speaking an abuse of notation) for the map is the usual function notation

$$F(p_1) = q_1, F(p_2) = q_2, ... \tag{7.10}$$

If $s \in P$ is in the first slot of some ordered pair, then s cannot be in the first slot of any other ordered pair of the map (collection) F. Alternatively, if F(s) = q, then it is not allowed that F(s)=q', $q \neq q'$. F(s) can map to one and only one element of Q.

In addition, if $t \in P$ is not in the first slot of some ordered pair of F, then the map is invalid. Alternatively, F(t) **must** map to some element of Q for every $t \in P$.

Taking $F : P \to Q$ as a set of ordered pairs, if P is finite, then

$$|F| = |P| \tag{7.11}$$

where |F| is the cardinality of the set of ordered pairs. If P is finite, the number of ordered pairs must equal the number of elements of P. That is, the cardinality (size) of the collection of ordered pairs defining the map F equals the cardinality (size) of P, since every element of P must be in the first slot of one and only one ordered pair. We will see later how the analog of this is defined for the usual function notation.

Different maps may be defined between the same two sets. Consider

$$\begin{aligned} &A = \{1, 2, 3\}, \quad B = \{a, b, c, d, e\} \\ &F : A \to B \end{aligned} \tag{7.12}$$

We define different versions of map F.

version 1 ================================

$$F = \{(1, b), (2, d), (3, a)\} \tag{7.13}$$

This map is illustrated in the following figure

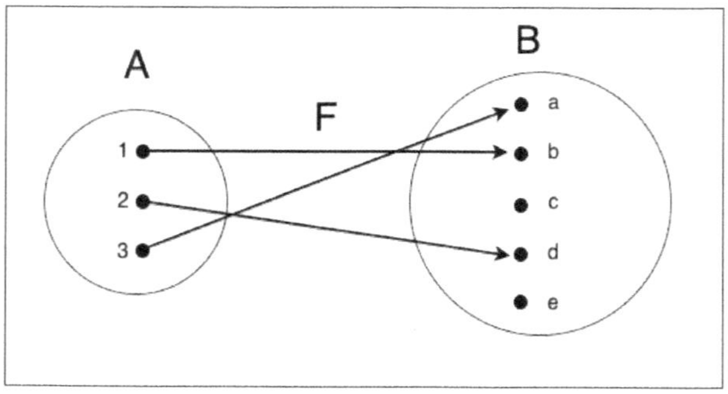

Figure 7.1:

Every element of A is in the first slot of some ordered pair, and is mapped to some element of B. Not every element of B is mapped from A by F.

In the usual function notation, the map F would be notated

$$F(1) = b, \quad F(2) = d, \quad F(3) = a \tag{7.14}$$

This map is said to be **one-to-one (1-1)** because, for every $s, s' \in A, s \neq s'$, then $F(s) \neq F(s')$.

A different map is F(1) = a, F(2) = e, F(3) = c. This is also a 1-1 map. (exercise: write this map as a set of ordered pairs.)

version 2 ==============================

$$
\begin{aligned}
F &= \{(1, a), (2, a), (3, e)\} \\
F(1) &= F(2) = a, F(3) = e
\end{aligned}
\tag{7.15}
$$

illustrated by the following figure

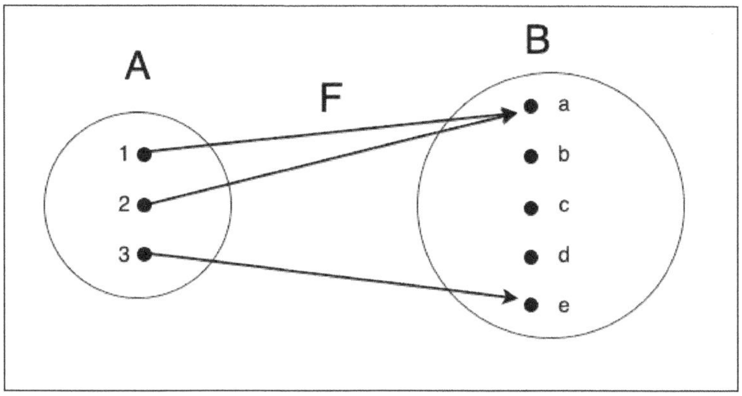

Figure 7.2:

In this example, elements 1, 2 of A are mapped to the **same** element 'a' of B. This is a perfectly legitimate map. F is not 1-1 because two elements of A map to the same element of B.

version 3 ==============================

$$
\begin{aligned}
F(1) &= F(2) = F(3) = e \\
F &= \{(1, e), (2, e), (3, e)\}
\end{aligned}
\tag{7.16}
$$

In this example, the map F is the **constant function**. It is not 1-1.

==============================

The following is a different sort of map, using

$$
A = \{1, 2, 3\}, \quad C = \{a, b, c\}
\tag{7.17}
$$

and map $F : A \to C$ such that

$$
\begin{aligned}
&F(1) = c, \quad F(2) = b, \quad F(3) = a \\
&F = \{(1, c), (2, b), (3, a)\}
\end{aligned}
\tag{7.18}
$$

In this case, F is 1-1 and also every element of C is included in the map. This map is illustrated below.

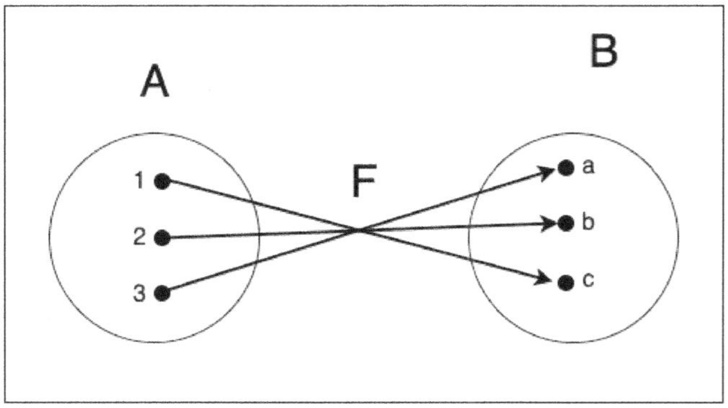

Figure 7.3:

One useful aspect of function notation is that the argument to the function as well as the value of the function may be a set. Using the example of equation 7.14, we may write

$$F(\{1,2\}) = \{b,d\}, \quad F(\{3\}) = \{a\}, \quad F(\{1,2,3\}) = \{a,b,d\} \tag{7.19}$$

For equation 7.15

$$F(\{1\}) = F(\{1,2\}) = F(\{2\}) = \{a\}, \quad F(\{3\}) = \{e\}, \quad F(\{1,2,3\}) = \{a,e\} \tag{7.20}$$

Note in this example the element 'a' appears at most once in the sets on the right side of each expression (no repeats in sets).

7.3 domain, codomain, image, range

Consider a map $F : A \to B$ from some set A to some set B. The set A is called the **domain** of the map. The set B is called the **codomain**.

The set **Im(F)**

$$Im(F) = \bigcup_{a \in A} \{F(a)\} = \bigcup_{a \in A} F(\{a\}) = F(A) \subseteq B \tag{7.21}$$

is the **image** of F. The image of F is the set of all elements of B which are actually mapped from A by F, and is illustrated below. (In equation 7.21, F(a) is not a set, while $F(\{a\})$ and $\{F(a)\}$ are sets. Union applies only to sets.)

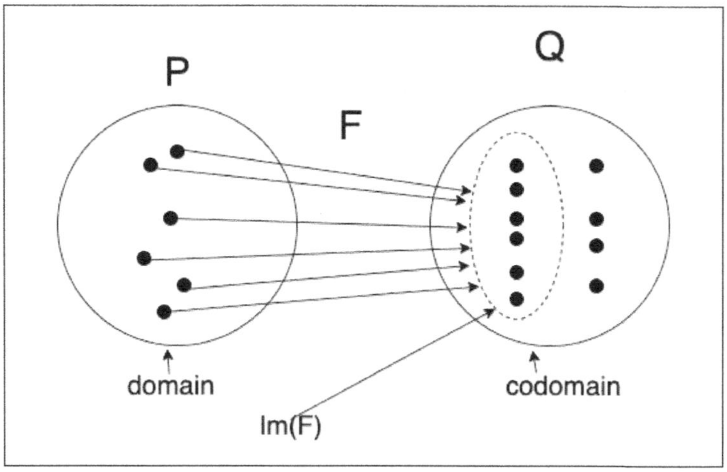

Figure 7.4:

The notation F(A), as in equation 7.21, is common. The image of F is also called the **range**, a term which seems to be no longer in favor. The image of F is a subset of the codomain, and it is possible that the image equals its codomain, as in equation 7.18, which is $F(\{1,2,3\}) = F(A) = C = \{a,b,c\}$.

The box indicates a use of the terminology of this section.

Each element of the **domain** is mapped once and only once to some element of the **image**. Two or more distinct elements of the domain can be mapped to the same element of the image.

7.4 injective, surjective, bijective maps

These three definitions are commonly used to differentiate important types of maps. The three categories correspond to three significant mathematical/set-theoretical situations. It's best to bite the bullet and, even if boring, learn the definitions.

Definition 7.2

Let P,Q be any non-empty sets, and $F : P \to Q$ be a map, with $p_i \in P$, q_i inQ.

*The map is **injective** if there do not exist two distinct elements p_i, p_j of P such that $F(p_i) = F(p_j)$. An injective map is called 1-1.*

*The map is **surjective** if, for each $q_i \in Q$, there exists a $p_j \in P$ such that $F(p_j) = q_i$. That is, $Im(F) = Q$ $(F(P) = Q)$.*

*The map is **bijective** it if is both injective and surjective.*

*For a bijective map, the elements of the sets P and Q are said to be in **1-to-1 correspondence**.*

A general map may be injective, surjective, both (bijective) or none of these.

If a map $F : P \to Q$ is surjective or bijective, $Im(F) = F(P) = Q$. If a map is not surjective, $Im(F) = F(P) \subset Q$, proper subset.

An arbitrary map may not fall into one of these 3 categories. The map, equation 7.15, is neither injective nor surjective (and therefore not bijective). It is not injective because two elements of A (1 and 2) are mapped to the same element (a) of B. It is not surjective because there is no element of A which is mapped to "b" (for example).

See figure below, which is valid for *finite sets*.

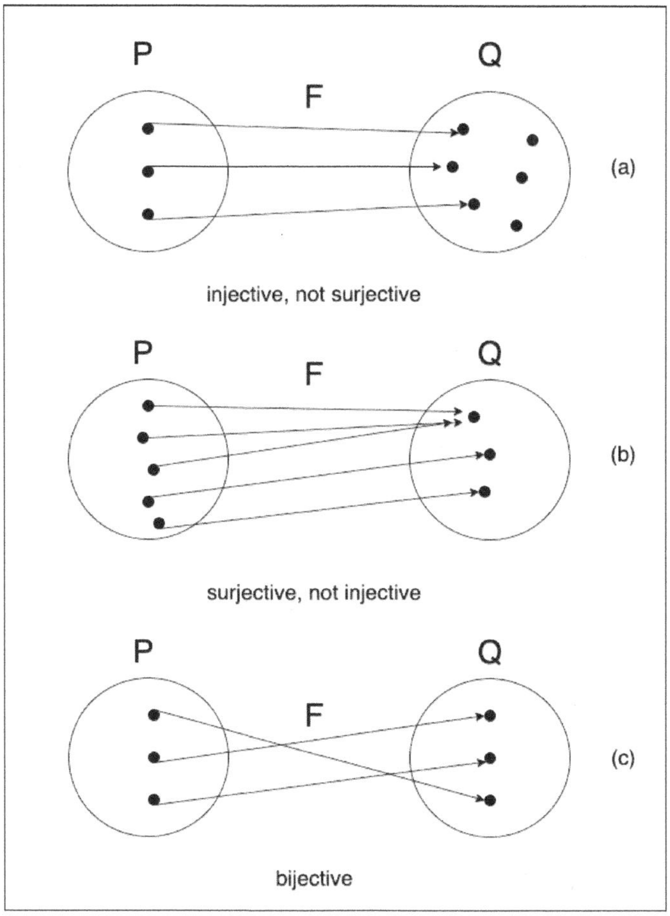

Figure 7.5:

$$(a)\ Im(F) \subset Q, \quad |P| < |Q|$$
$$(b)\ Im(F) = Q, \quad |P| > |Q|$$
$$(c)\ Im(F) = Q, \quad |P| = |Q|$$

- The map of equation 7.14 is injective because F(1), F(2) and F(3) are all different. It is not surjective.

- The map of equation 7.15 is *not* injective because F(1) = F(2). It is also not surjective.

- Similarly the map of equation 7.16 is neither injective nor surjective.

- The map of equation 7.18 is injective and surjective, so it is bijective.

Theorem 7.3

For the map $F : P \to Q$, with both P, Q finite:
If $|P| > |Q|$ then F cannot be injective. It could be surjective.
if $|P| < |Q|$ then F cannot be surjective. It could be injective.
If $|P|=|Q|$ and F is injective then F is bijective
If $|P|=|Q|$ and F is surjective, then F is bijective.

Exercise: Try to prove this theorem before reading further.

Suppose |P| > |Q|. If F is injective, then each element of P must be mapped to a different element of Q, but there are not enough elements of Q to go around. In this case, F could be (but need not be) surjective.

Suppose |P| < |Q|. If F were surjective, then each element of Q must be the target of a map from at least one element of P, but there are not enough elements of P to do this. In this case, F could be (but need not be) injective.

If |P| = |Q| and F is injective. If F were not surjective, then at least two elements of P must be mapped to the same element of Q. But that means F is not injective, a contradiction. Therefore F is surjective, and since F is assumed injective, then F is bijective.

A similar argument holds if |P| = |Q| and F is surjective.

The following examples should be understood. Let

$$A = \{a, b\} \tag{7.22}$$

======================

Here is a toy example

$$\begin{aligned} &F, G : A \to \mathbb{Z}_+ \\ &F(a) = 27, \quad F(b) = 433 \\ &G(a) = 1, \quad G(b) = 1 \end{aligned} \tag{7.23}$$

F is injective but not surjective. G is neither injective nor surjective, and is the **constant function**.

======================

This next is actually realistic.

$$\begin{aligned} &F, G : \mathbb{Z}_+ \to A \\ &F(n) = a \;\; \forall n \in \mathbb{Z}_+ \\ &G(n) = \begin{cases} a & n\,even \\ b & n\,odd \end{cases} \end{aligned} \tag{7.24}$$

F is neither surjective nor injective and is the **constant function**. G is surjective but not injective (G(2)=G(4)=G(6)=.... = a) and appears in equivalence relations and in topology.

======================

The next map will be seen to be quite important. Let

$$EV = \{2, 4, 6, 8, ...\} \tag{7.25}$$

the infinite set of even integers.

$$\begin{aligned} &F : \mathbb{Z}_+ \rightarrow EV \\ &F(n) = 2n \end{aligned} \tag{7.26}$$

This function is both surjective and injective, and hence is bijective. It is visualized in figure 6.1.

======================

This next map is surjective but not injective.

$$\begin{aligned} &F : \mathbb{Z}_+ \rightarrow \mathbb{Z}_+ \\ &F(n) = \begin{cases} \frac{n+1}{2} & n\ odd \\ \frac{n}{2} & n\ even \end{cases} \end{aligned} \tag{7.27}$$

For example, $F(1) = F(2) = 1, F(3) = F(4) = 2, F(5) = F(6) = 3, ...$

======================

This next is injective but not surjective.

$$\begin{aligned} &F : EV \rightarrow \mathbb{Z}_+ \\ &F(n) = n \end{aligned} \tag{7.28}$$

$F(2) = 2, F(4) = 4, F(6) = 6, ...$

7.5 pre-image

For the map $F : P \rightarrow Q$, there is a reverse operation notated F^{-1}. **This operation is in general not a map (that is, it is not a function). It is a function only in certain circumstances.** It is defined as follows:

Definition 7.4

*For map $F : P \to Q$, of non-empty sets P, Q, the **pre-image** $F^{-1}(q)$ of some element $q \in Q$ is a **subset** of P where*

$$F^{-1}(q) = \{p|p \in P \ \& \ F(p) = q\} \subseteq P \tag{7.29}$$

The pre-image of a subset $W \subseteq Q$ is the set $F^{-1}(W)$

$$F^{-1}(W) = \left(\bigcup_{x \in W} F^{-1}(x) \right) \subseteq P \tag{7.30}$$

A pre-image is a set. Equation 7.29 reads: The pre-image of an element $q \in Q$ is a subset of P, notated $F^{-1}(q)$. This pre-image contains all $p_1, p_2, ... \in P$ such that $F(p_1) = F(p_2) = ... = q$.

Here is a figure for both equations above.

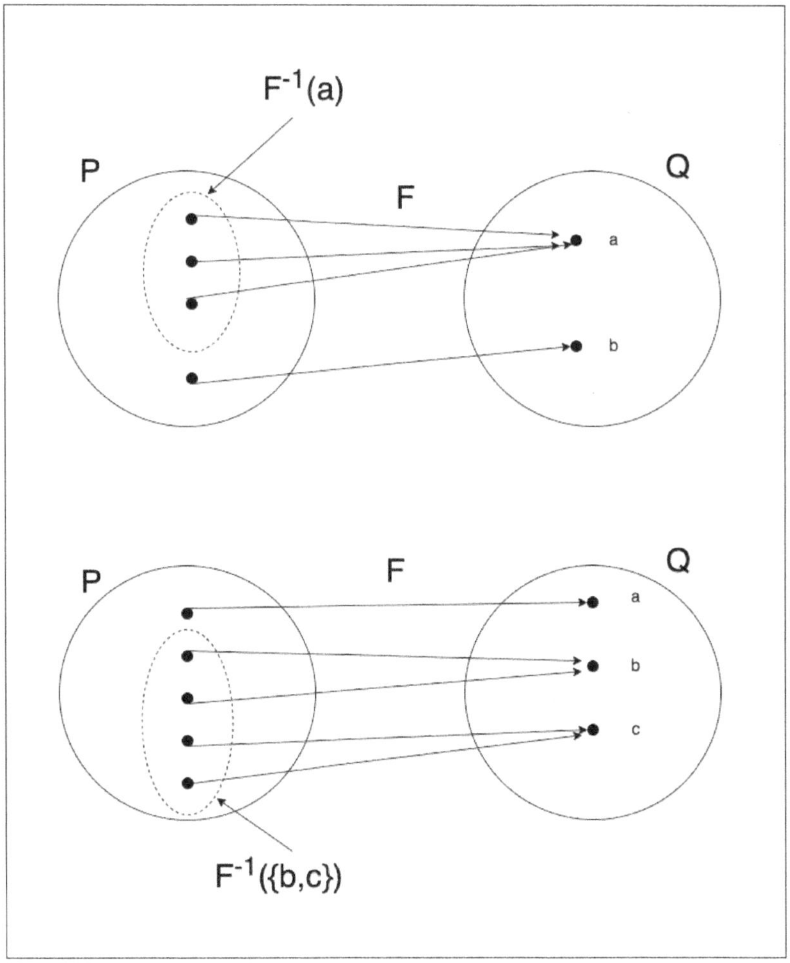

Figure 7.6:

This figure shows two *different* maps, both called F, and displays a pre-image for each map. In the top half of the figure, we see that $F^{-1}(a)$ is a set containing three elements of P. That is, F^{-1} 'sends' a to three elements of P. As discussed above, one critical property of a function is that it maps a given element of the domain to one and only one element of the codomain. If F^{-1} were a genuine function, then Q would be its domain, P its codomain and 'a' in Q should be 'mapped' to only a single element of P. But since the pre-image of 'a' consists of three elements of P, then $\mathbf{F^{-1}}$ **cannot be a function**. Similar considerations holds for the map in the bottom half of the above figure. F^{-1} is a pre-image, but not a function.

For another example, from equation 7.12, $A = \{1, 2, 3\}$, $B = \{a, b, c, d, e\}$ and map F of equation 7.15, which is

$$F(1) = F(2) = a, \quad F(3) = e \tag{7.31}$$

We see that

$$F^{-1}(a) = \{1, 2\} \subset A, \quad F^{-1}(e) = \{3\} \subset A \tag{7.32}$$

Therefore F^{-1} is a pre-image but not a function.

Continuing this example, what about $F^{-1}(b), F^{-1}(c), etc$? There is no $x \in A$ such that $F(x) = b$ *or* c *or* d. This is not covered in definition 7.4, so that definition must be enhanced.

Definition 7.5

For map $F : P \to Q$*, if for some* $q \in Q$ *there is* **no** $p \in P$ *such that* $F(p) = q$*, then*

$$F^{-1}(q) = \emptyset \subset P \tag{7.33}$$

The pre-image is the empty set.

For set $W \in Q$*, the second part of definition 7.4, equation 7.30, is still valid.*

For the previous example, for the elements b,c,d of B

$$F^{-1}(b) = \emptyset, \quad F^{-1}(c) = \emptyset, \quad F^{-1}(d) = \emptyset, \quad F^{-1}(\{b, c, d\}) = \emptyset \tag{7.34}$$

See how useful it is to have the empty set be a subset of any set, so the right hand sides of definition 7.4 carry over to definition 7.5. It is always the case that $\emptyset \subset P$ (theorem 4.9). **The pre-image of a function F is *always* a subset (not necessarily proper) of the domain of F.**

Exercise: describe a map $F : P \to Q$ such that $F^{-1}(Q) = P$.

The intermediate case is also covered by definition 7.5

$$F^{-1}(\{a, d\}) = F^{-1}(a) \cup F^{-1}(d) = \{1, 2\} \cup \emptyset = \{1, 2\} \tag{7.35}$$

Consider the example equation 7.24. For the *first* case, $F^{-1}(a) = \mathbb{Z}_+, \quad , F^{-1}(b) = \emptyset$. For the *second* case, $F^{-1}(a) = EV, \; F^{-1}(b) = \{1, 3, 5, 7, ..., \}$.

For the example equation 7.28, if T is **any** set of odd numbers or any individual odd number, $F^{-1}(T) = \emptyset$. For any even number k, $F^{-1}(k) = \{k\}$. Also, for example, $F^{-1}(\{2, 6, 22, 88\}) = \{2, 6, 22, 88\}$ The intermediate case is straightforward, $F^{-1}(\{2, 4, 5\}) = \{2, 4\}$.

7.6 surjective maps and partitions.

This section logically belongs here and provides insight into surjective maps. However, it can be skipped on a first reading. Come back to it before starting the section on equivalence relations.

There is a very close relation between surjective maps and partitions.

Definition 7.6

*A **partition** of any non-empty set S is a collection P of nonempty subsets of S such that each element of S is in one and only one subset.*

A partition P of S is such that $S = \bigcup P$.

$$\bigcup P = \bigcup_{p \in P} p = S \tag{7.36}$$

A partition partitions a set into pairwise disjoint sets. That is, for partition P, if $|P| \neq 1$ and if $A, B \in P$ are sets in P, $A, B \subset S$, then $A \cap B = \emptyset$.

The last part of this definition is relevant for functions, as will be seen.

Here are some examples. Let

$$S = \{1, 2, 3, 4, ..., 12\} \tag{7.37}$$

==================================

One partition of S is

$$P_1 = \{\ \{1, 3, 7, 9, 12\},\ \{2, 4, 5, 6, 8, 10, 11\}\ \} \tag{7.38}$$

Note that $\cup P_1 = S$. Note that $\{1, 3, 7, 9, 12\} \cap \{2, 4, 5, 6, 8, 10, 11\} = \emptyset$.

==================================

Another partition of S.

$$P_2 = \{\ \{3, 8, 10\ \}\ \{2, 5, 7, 11\},\ \{1, 4, 6, 9, 12\}\ \} \tag{7.39}$$

$\cup P_2 = S$. The sets of P_2 are pairwise disjoint.

==================================

There are two **trivial partitions** of S:

$$P_3 = \{\{1\}, \{2\}, \{3\}, ..., \{12\}\}, \quad P_4 = \{\ \{1, 2, 3, 4....12\}\ \} \tag{7.40}$$

We have $\cup P_3 = \cup P_4 = S$.

=====================================

Here are two partitions of $\mathbb{Z}_+$:

$$\begin{aligned} P_5 &= \{\ \{2, 4, 6, 8, ...\},\ \{1, 3, 5, 7, ...\}\ \} \\ P_6 &= \{\ \{1, 4, 7, 10, 13, ...\},\ \{2, 5, 8, 11, 14, ...\},\ \{3, 6, 9, 12, 15, ...\}\ \} \end{aligned} \tag{7.41}$$

$\cup P_5 = \cup P_6 = \mathbb{Z}_+$. P_5 partitions the integers into even and odd integers.

=====================================

Here is a partition of $\mathbb{Z}$, the set of all integers.

$$P_7 = \{\ \{0\},\ \{2, 4, 6, 8, ...\},\ \{1, 3, 5, 7, ...\},\ \{-2, -4, -6, -8, ...\},\ \{-1, -3, -5, -7\}\ \} \tag{7.42}$$

Partitions are intimately related to surjective maps, as follows.

Theorem 7.7

Let S, T be nonempty sets and $F : S \to T$ be a surjective map. Then the set

$$P = \{F^{-1}(t)\,|\,\forall t \in T\} \tag{7.43}$$

is a ***partition of S****. For $t, t' \in T,\ t \neq t'$ the pre-images are disjoint*

$$F^{-1}(t) \cap F^{-1}(t') = \emptyset \tag{7.44}$$

This means the sets of the partition P are pairwise disjoint, as required.

This theorem is unpacked below. If S is finite, then $|S| \geq |T|$ is required in order F to be surjective (see theorem 7.3). Because F is surjective, there is no $t \in T$ such that $F^{-1}(t) = \emptyset$. That is, the last part of definition 7.6 is not necessarily valid if F is not surjective (you should understand this. Draw a diagram, and recall that $\emptyset = \emptyset$).

Suppose the pre-images of a surjective function were not disjoint. Suppose one element $s \in S$ is in the pre-images of two different elements of T. This means that there must exist $t, t' \in T, t \neq t'$ such that $s \in F^{-1}(t)$, $s \in F^{-1}(t')$. Then, by definition of pre-image, we must have $F(s) = t$ and $F(s) = t'$, but since $t \neq t'$, such is not valid.

Consider the partition P_1, equation 7.38. The partition P_1 results from the map $F : S \to T = \{a, b\}$, where

$$F(\{1,3,7,9,12\}) = a, \quad F(\{2,4,5,6,8,10,11\}) = b \tag{7.45}$$

as seen in the figure below.

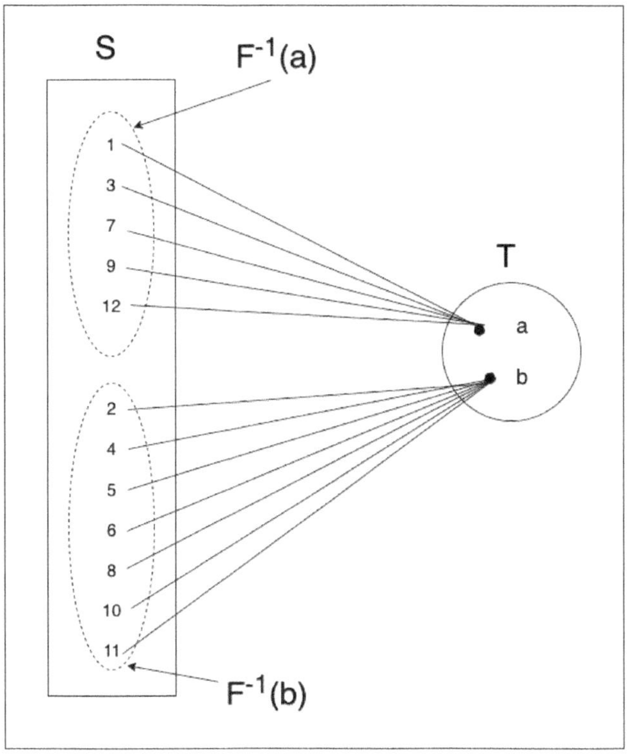

Figure 7.7:

The associated partition of S is

$$P_1 = \{F^{-1}(a), F^{-1}(b)\} \tag{7.46}$$

The partition P_2 results from the map $F : S \to T = \{a, b, c\}$, where

$$F(\{3,8,10\}) = a, \quad F(\{2,5,7,11\} = b, \quad F(\{1,4,6,9,12\}) = c \tag{7.47}$$

as in the figure below.

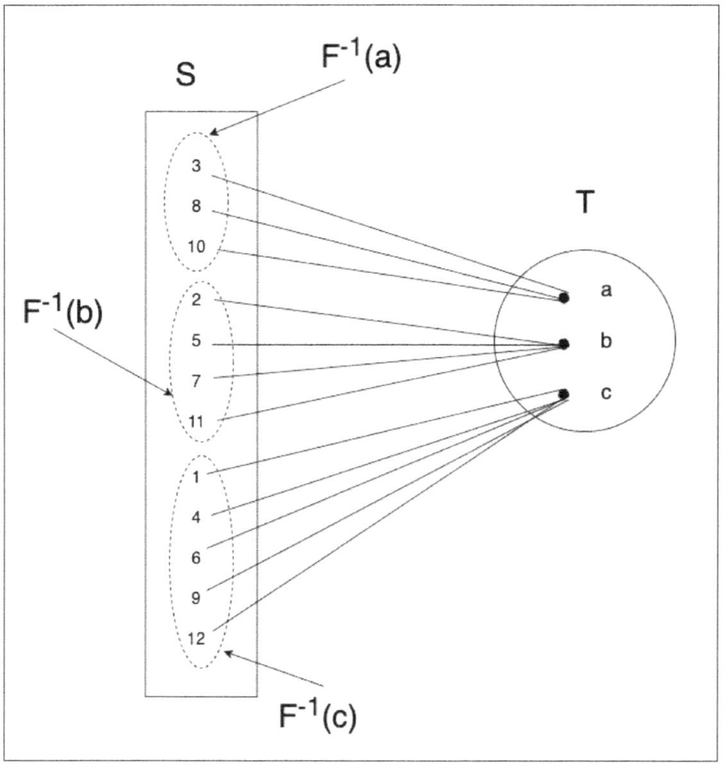

Figure 7.8:

The associated partition of S is

$$P_2 = \{F^{-1}(a), F^{-1}(b), F^{-1}(c)\} \tag{7.48}$$

The partition P_4, equation 7.40, results from a set T where $|T| = 1$, so F is the constant function. $F : S \to T,\ T = \{a\},\ F(1) = F(2)... = F(12) = a$.

The partition P_3, equation 7.40, results if F is a bijection.
$F : S \to T, T = \{x_1, x_2, ..., x_{12}\},\ F(1) = x_1, F(2) = x_2, ..., F(12) = x_{12}$.

The partition P_6 , equation 7.41, results from the map $F : \mathbb{Z}_+ \to \{a, b, c\}$,

$$F(\{1, 4, 7, 10, ...\}) = a, \quad F(\{2, 5, 8, 11, ...\}) = b, \quad F(\{3, 6, 9, 12, ...\}) = c \tag{7.49}$$

as in the figure below.

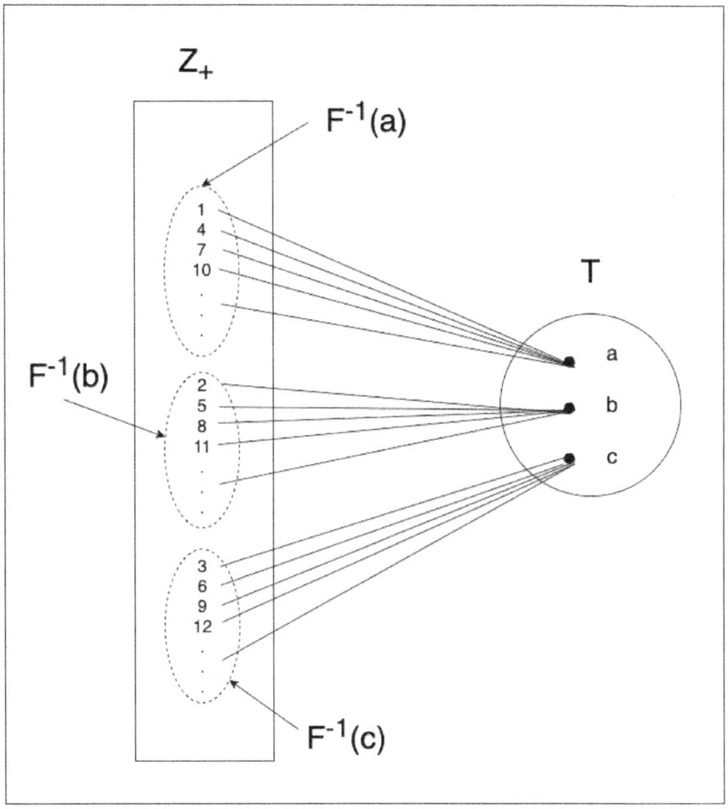

Figure 7.9:

The associated partition of Z_+ is

$$P_6 = \{F^{-1}(a), F^{-1}(b), F^{-1}(c)\} \tag{7.50}$$

7.7 injective maps and cardinality

The following is an important property of injective (1-1) maps.

Theorem 7.8

*Let A, B be **finite** sets and $F : A \to B$. If F is injective then $|A| \leq |B|$.*

This is a restatement of part of theorem 7.3.

The existence of an injective map may be used to determine whether the cardinality of one finite set is smaller or larger than the cardinality of another finite set. This will be important when we consider cardinal numbers..

7.8 bijective maps and inverses

Bijective maps are crucial to analyzing infinite sets.

Definition 7.9 *Let $F : P \to Q$ be a bijective map. In this particular case, the notation F^{-1} is the **inverse** of F and it is a 1-1 surjective **map***

$$F^{-1} : Q \to P \tag{7.51}$$

If $p \in P, q \in Q$ such that $F(p) = q$, then

$$F^{-1}(q) = p \tag{7.52}$$

***The inverse F^{-1} is also a bijective map.** The inverse of an element of Q is not a set but is an element of P.*

It is unfortunate that the inverse and the pre-image have the same notation.

This means that F^{-1} is a one-to-one function, so for any two $q, q' \in Q$, we must have $F^{-1}(q) \neq F^{-1}(q')$. This is obvious in the figures 7.5 and 6.1, and in equations 7.18 and 7.26.

The inverse is not a pre-image. It is a map (a function) and, if its argument is a single element, it is not a set. (Like any function if its argument is a set, then it is a set.)

Here is yet another example. Let

$$A = \{1, 4, 9, 12\}, \quad B = \{a, r, s, m\} \tag{7.53}$$

Let $F : A \to B$, such that

$$F(1) = r, \quad F(4) = a, \quad F(9) = m, \quad F(12) = s \tag{7.54}$$

F is a bijective map. Therefore F^{-1} is a bijective map.

$$F^{-1}(r) = 1, \quad , F^{-1}(a) = 4, \quad F^{-1}(m) = 9, \quad F^{-1}(s) = 12 \tag{7.55}$$

Note that $F^{-1}(B) = A$.

Perhaps this will be useful.

Consider a bijective map $F : P \to Q$.

There is **no** $q \in Q$ such that $F^{-1}(q) = \emptyset$.
If $F(p) = q$, there is **no** $p' \in P$, $p' \neq p$, such that $F^{-1}(q) = p'$.

This is just another way of saying that F is both injective and surjective.

The following theorem will be useful.

Theorem 7.10

*P, Q are nonempty sets, F a map $F : P \to Q$. The map F is bijective if and only if **all** of the following are true for all $p \in P$ and all $q \in Q$:*

$$\begin{aligned} F(F^{-1}(q)) &= q \\ F^{-1}(F(p)) &= p \\ F(F^{-1}(Q)) &= Q, \\ F^{-1}(F(P)) &= P \end{aligned} \tag{7.56}$$

For the third case, $F^{-1}(Q) = P$, and then $F(P) = Q$. The fourth case is just the reverse. $F(P) = Q$ and then $F^{-1}(Q) = P$. The first two cases are the same but with particular elements p, q.

7.9 isomorphisms

The concept of an isomorphism is a very important one. There are group isomorphisms (called homomorphisms), topological isomorphisms (called homeomorphisms), and so forth. A **set isomorphism** is the simplest of these, and the concept is central to much of the material in this book.

Definition 7.11 *Two non-empty sets P, Q are* ***isomorphic*** *or* ***set isomorphic*** *if there is* ***some*** *bijection F from one to the other. An isomorphism between two sets is notated*

$$P \cong Q \tag{7.57}$$

Note that if $P \cong Q$ then $Q \cong P$, because the inverse F^{-1} is also a bijective map.

The nature of the bijection is not specified. Some bijection just has to exist. **Any bijection between two sets means the sets are isomorphic.**

When two sets are isomorphic it means that certain properties of the sets are the same. At this point, as will be seen, it is the cardinality of isomorphic finite sets that is the same. (For infinite sets, cardinality is not the same concept as size).But there are other properties as well that will be similar.

If two sets are equal, they are isomorphic. However, isomorphic sets need not be equal.

For example the sets $A = \{1, 2, 3\}, B = \{a, b, c\}$ are isomorphic but they are not equal. There are a number of possible bijections between the sets. For example, $1 \to a, 2 \to b, 3 \to c$ is one bijection. Another is $1 \to c, 2 \to a, 3 \to b$.

The following is a useful theorem. We state it here. For finite sets, this can be proved using theorem 7.3. For infinite sets, the proof will be given later, using function composition, introduced later.

Theorem 7.12 *Given 3 sets P, Q, R. If $P \cong Q$ and $Q \cong R$ then $P \cong R$.*

In other words, two sets which are isomorphic to a third set are isomorphic to each other.

8 countably infinite sets

8.1 countable finite set and cardinality

Theorem 8.1 *Let Q be a finite non-empty set and define*

$$Z_N = \{1, 2, 3, ..., N\} \tag{8.1}$$

the set of positive integers from 1 through N. If there exists a bijective map $F : Z_N \to Q$*, so that* Z_N *and Q are isomorphic, then the* ***cardinality*** *of Q is*

$$|Q| = N \tag{8.2}$$

That is, for a finite set Q, the cardinality of Q equals the number of elements of the finite set of integers isomorphic to Q (same number of elements in the two sets).

Consider the figure

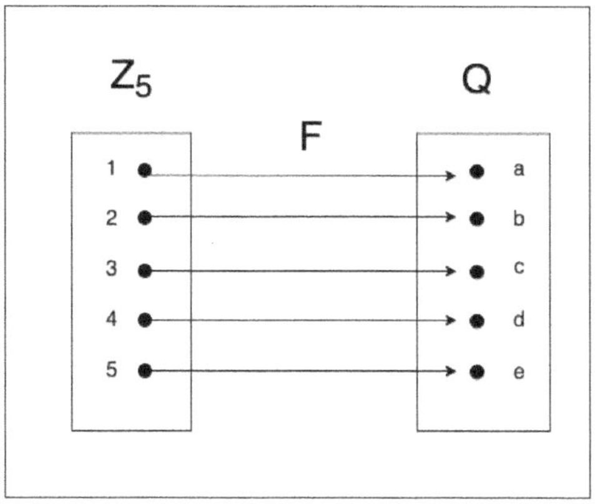

Figure 8.1:

F is a bijection and it should be clear that Z_5 and Q are isomorphic and have the same cardinality.

Definition 8.2

Because a finite set is always isomorphic to some finite set of positive integers, the finite set is said to be ***countable****. Since we can count the elements of the set using integers (as a proxy), the finite set can be counted.*

8.2 countably infinite set and cardinality

The size, or cardinality, of an infinite set is, at this point, not defined. Cardinality is a number, but ∞ is not as number and hence cannot be a cardinality. So what is the cardinality of Z_+? We define a **symbol** for it: $\aleph_0$.

Definition 8.3

The cardinality of Z_+, the set of positive integers, is ***defined*** *to be $\aleph_0$ (pronounced aleph-zero). The set Z_+ is said to be* ***countably infinite****.*

As we will see, the symbol $\aleph_0$ can, in certain circumstances, be manipulated as if it were a number. That is, an arithmetic can be defined where the symbol $\aleph_0$ can be added, subtracted and multiplied to itself and to other cardinalities. For now, just accept that the cardinality of Z_+ is $\aleph_0$.

Continuing from section 6, Z_+ is the (infinite) set of positive integers and define

$$EV = \{2, 4, 6, ...\} \tag{8.3}$$

the infinite set of even integers.

We may define a map

$$\begin{aligned}
&F : EV \to Z_+ \\
&F(n) = \frac{n}{2} \\
&F(2) = 1, \quad F(4) = 2, \quad F(6) = 3, \quad .... \\
&F^{-1} : \mathbb{Z}_+ \to EV \\
&F^{-1}(n) = 2n \\
&F^{-1}(1) = 2, \quad F^{-1}(2) = 4, ...
\end{aligned} \tag{8.4}$$

The map F is **bijective** and F^{-1} is the inverse map. The sets Z_+ and EV are isomorphic - they are in 1-to-1 correspondence. In section 8.1 we have defined the cardinality of a finite set as N if it is in 1-to-1 correspondence with the set of N integers. This definition of cardinality can be extended to the case of infinite sets.

The idea is this: **the cardinality of the set of even integers is the same as the cardinality of Z_+ because there is a bijective map, equation 8.4, between the two sets.**

We now generalize the concept of cardinality from the case of finite sets to the case of the infinite sets. For a finite set S, cardinality has been redefined to mean bijection from S to to a finite set of integers. The number of elements in the finite set of integers is the *cardinality of S*. This is a transfer from the set S of interest to a set of integers, by means of a bijective map.

We now define cardinality of an infinite set by generalizing this procedure.

Definition 8.4

Any set which is isomorphic to (has bijection to) Z_+ is ***countably infinite****. Its cardinality is $\aleph_0$.*

The term *countably infinite* has a similar meaning for an infinite set of integers as the term *countable* does for as finite set. The elements can be *counted* (they are integers after all), even though there are an infinite number of them.

Many unequal sets have cardinality $\aleph_0$. Along with $\mathbb{Z}_+$, the set EV, equation 8.4 has cardinality $\aleph_0$. The set J, equation 6.14, has cardinality $\aleph_0$. More examples are given below.

The quantity $\aleph_0$ is a symbol - it is not any finite number. Because it is a kind of size, it is called a **cardinality**, and sometimes a **cardinal number**. Its meaning is determined by the way it is used. One can think of it as the smallest number which is larger than any finite number. At present, this sentence should make no sense whatsoever. However, it turns out that this statement does make sense (see the section *ordinal numbers.*).

The following theorem is stated in two equivalent parts. We put off the proof until we study the composition of functions, but it is useful to state it here.

Theorem 8.5

Given two sets A, B, if A is countably infinite (isomorphic to $\mathbb{Z}_+$) and if B is isomorphic to A, then B is countably infinite as well (it is also isomorphic to $\mathbb{Z}_+$).

If any two sets are countably infinite, they are isomorphic.

Let

$$A = \{a_1, a_2, a_3, ...\}, \quad B = \{b_1, b_2, b_3, ...\} \tag{8.5}$$

Since A is isomorphic to $\mathbb{Z}_+$, there exists some bijection $F(a_i) = i$. Since B is isomorphic to A, there exists some bijection $G(b_i) = a_i$. Therefore $F(G(b_i)) = F(a_i) = i$ is the bijection from B to $\mathbb{Z}_+$, and so B is countably infinite.

This is an example of *function composition*, which will be discussed later.

Recall the set of positive integers where 8 has been removed.

$$A = \{1, 2, 3, 4, 5, 6, 7, \;\; 9, 10, 11, 12, ...\} \tag{8.6}$$

We can now understand in detail why this set is countably infinite. There is a bijection between the positive integers and the set A.:

$$\begin{aligned} &F : \mathbb{Z}_+ \to A \\ &F(1) = 1, \;\; F(2) = 2, ..., F(7) = 7, \\ &F(8) = 9, \;\; F(9) = 10, \;\; F(10) = 11, ... \end{aligned} \tag{8.7}$$

The cardinality of set A is $\aleph_0$.

Consider the infinite set of squared integers

$$SQ = \{1, 4, 9, 16, 25, ...\}$$

There is a bijective map

$$\begin{aligned} &F : SQ \to Z_+ \\ &F(n) = \sqrt{n} \\ &F(1) = 1, \quad F(4) = 2, \quad F(9) = 3, \quad .... \end{aligned}$$

The infinite set of squared integers is countably infinite, and its cardinality is $\aleph_0$.

The following sets are also countably infinite, each with cardinality $\aleph_0$.

$$\begin{aligned} &\{1, \frac{1}{2}, \frac{1}{3}, \frac{1}{4}, ...\} \\ &\{\frac{1}{10}, \frac{1}{20}, \frac{1}{30}, ...\} \\ &\{a, aa, aaa, aaaa, aaaaa, aaaaaa, ...\} \\ &\{(3, x), (6, x), (9, x), (12, x), ...\} \end{aligned} \tag{8.8}$$

Any pair of the sets, equation 8.8, are isomorphic, and each is isomorphic to $\mathbb{Z}_+$. It would be good practice to explicitly write the bijection between $\mathbb{Z}_+$ and each of the above sets. For the sets with sequences of 'a', use an abbreviation, for example, $a^5 = aaaaa$, $a^9 = aaaaaaaaa$ and so forth. Also, make up some of your own countably infinite sets - the weirder the better.

Obviously there are an infinite number of countably infinite sets, every one of which has exactly the same cardinality $\aleph_0$.

> The cardinality of a finite set is an ordinary integer, the number of elements. The cardinality of a countably infinite set is a symbol $\aleph_0$.

We have resolved the paradox about the size of infinite sets, because we have redefined size as cardinality. Let P, Q be any two countably infinite sets. Then $P \cong Q$. Suppose *also* that $P \subset Q$. Then P is a proper subset of Q. (The set of integers divisible by 4 is a proper subset of the set of even integers, and both sets are countably infinite.) Nonetheless P and Q have the same cardinality because they are isomorphic.

We have *resolved the paradox of size* by completely redefining what is meant by the **size** of any set. We call it **cardinality**. Admittedly this is a little (a lot?) like cheating. Nonetheless we have defined cardinality in analogy with the cardinality of finite sets, **the cardinality of both finite and countably infinite sets being determined by isomorphisms with the appropriate set of positive integers**.

More importantly is the method used to define the cardinality of an infinite set and relate it to the cardinality of finite sets. We have made an end run around explicitly counting elements by using a map - in this case a bijective map. The procedure is:

1. Define the cardinality of a finite set S in terms of a bijective map from the set S to a finite set of positive integers. (This replaces counting elements, although by itself this is not particularly interesting.)
2. Generalize the above to infinite sets by a bijective map from the set to $\mathbb{Z}_+$.

We will see that other properties of infinite sets can be defined by a similar procedure but with different types of maps.

The following theorem is useful.

Theorem 8.6

Let A be a countably infinite set and let $B \subset A$ be an infinite set and a proper subset of A. Then B is countably infinite.

In other words, an infinite proper subset of a countably infinite set is countably infinite.

The proof is a good exercise.

8.3 the set of rational numbers

8.3.1 rational numbers

A rational number is the quotient of two integers. The following are rational numbers:

$$1 = \frac{1}{1}, \quad 8 = \frac{8}{1}, \quad \frac{1}{8}, \quad \frac{2}{7}, \quad \frac{962}{29}, \frac{289645134278}{92887652423098162} \tag{8.9}$$

Considering the rational number as a decimal $n_1 n_2 .... \; . \; x_1 x_2 x_3 ...$, the sequence of digits $x_1 x_2 ...$ after the decimal point **always eventually** becomes a repeating sequence, $\frac{1}{3} = 0.3333333$, the 3 repeats, $\frac{1}{2} = 0.50000000$, the 0 repeats. $\frac{7}{11} = 0.63636363.....$ Finally

$$\frac{29}{31} = 0.935487096774193548709677419354870967741935487096 7741..... \tag{8.10}$$

where 4193548709677 repeats.

Definition 8.7

$\mathbb{Q}_+$ is the set of all positive rational numbers.
$\mathbb{Q}$ is the set of all rational numbers, positive, negative, and 0.
$\mathbb{Q}^*$ is the set of all rational numbers except for 0.

The following are four important properties of rational numbers.

Theorem 8.8

The sum, difference, product and quotient of two rational numbers is a rational number.

squeeze theorem. *Between any two rational numbers is an infinite number of rational numbers.*

The set $\mathbb{Q}_+$ is countably infinite.

The sets $\mathbb{Q}$, $\mathbb{Q}_+$ and $\mathbb{Q}^*$ are pairwise isomorphic.

The first is proved using arithmetic. For the second, if a/b and c/d are rational, where a,b,c,d are integers, then

$$x = \frac{1}{2}\left(\frac{a}{b} + \frac{c}{d}\right) = \frac{ad + bc}{2bd} \tag{8.11}$$

lies between the two rationals, and is the ratio of two integers and so is rational. Just iterate this and show that, for any step, there is a rational number, say r, between the one before and c/d. Then if there are only a finite number in the interval, there must be a largest r, and then show there is a larger one less than c/d.

I like to think of this property as the **squeeze** property. No matter how close two rational numbers are, a third rational number can always be *squeezed* between the two. The squeeze property distinguishes numbers from the reality of things. If two atoms are neighbors - in a crystal, say - whose positions are rational numbers - one cannot squeeze a third atom between without displacing one or both of the original two.

The third and fourth parts of the above theorem will be proved later.

Consider the set

$$A = \{0.3, 0.33, 0.333, 0.3333, ...\} = \{\frac{3}{10}, \frac{33}{100}, \frac{333}{1000}, ...\} \tag{8.12}$$

Using the notation $3_n = 333...3$, n times, we can write A in the form $A = \{\frac{3_n}{10^n} | n \in \mathbb{Z}_+\}$. **Is $\frac{1}{3}$ in A?**. The answer is no because, for any $n \in \mathbb{Z}_+$, $\frac{1}{3} \neq \frac{3_n}{10^n}$ and A contains only the numbers of the form $\frac{3_n}{10^n}$.

8.3.2 rational numbers and repeating termination

The content of this subsection is not used until we consider irrational numbers. But it is included here because this section is largely about rational numbers. We start with a definition, and then give examples and justification.

Definition 8.9

A number n, $0 \leq n < 1$, is rational iff its ***decimal representation*** *is one of the three forms*

$$\begin{aligned} n &= 0\ .\ x_1x_2, ...x_N0000000000000.......\quad x_N \neq 0 \\ n &= 0\ .\ x_1x_2, ...x_N9999999999999......\quad x_N \neq 9 \\ n &= 0\ .\ x_1x_2, ...x_Ny_1y_2...y_py_1y_2...y_py_1y_2...y_p..... \end{aligned} \tag{8.13}$$

In all cases, x_i, y_k are digits 0 - 9 and the integers N, p are finite.

In the second case, $n = 0\ .\ x_1x_2, ...x_{N-1}(x_N + 1)000000......$

The third case is called a ***repeating termination****, and is assumed not the same as the second case. Here, $p \geq 1$. In this case, the string of digits $x_1x_2, ...x_N$ is optional (for example, in the case of 1/3).*

Examples of the third case are given in section 8.3.1.

In the first case,

$$n = \sum_{i=1}^{N} \frac{x_i}{10^i}$$

which is the finite sum of rational numbers and so is rational.

The second case is

$$\begin{aligned} n &= \sum_{i=1}^{N} \frac{x_i}{10^i} + \sum_{i=N+1}^{\infty} \frac{9}{10^i} \\ &= \sum_{i=1}^{N} \frac{x_i}{10^i} + \frac{9}{10^N} \sum_{i=1}^{\infty} \frac{1}{10^i} \\ &= \sum_{i=1}^{N} \frac{x_i}{10^i} + \frac{1}{10^N} \\ &= \sum_{i=1}^{N-1} \frac{x_i}{10^i} + \frac{x_N + 1}{10^N} \rightarrow 0\ .\ x_1 x_2 ... x_{N-1}(x_N + 1) \end{aligned}$$

Therefore, **the second case is the same as the first where x_N is increased by 1**. This will turn out to be important.

For the third case, which holds if n is not either of the first two cases, we illustrate with an example. Let $n = 0.417212121....$, a repeating termination. Define z, $z = 10^3 n - 417 = 0.121212.....$ Because 0.121212.... is an infinite repeating sequence, we can see that $10^2 z = 12.121212.... = 12 + z$. Here we have used the fact that the sequence $z = 0.121212....$ is countably infinite. Therefore

$$\begin{aligned} 10^2 z = 12 + z \rightarrow z = \frac{12}{99} \\ n = \frac{z + 417}{10^3} \end{aligned}$$

z is rational and therefore n is rational.

The general case is a simple and tedious generalization of this example, and is left as an exercise.

Next we show that if $n = \frac{x}{y}$, $x < y$, x,y integers, 0 < n < 1, then the decimal representation of n has a repeating termination (if it is not all zeros). This can be shown using the grade-school long division algorithm . Consider the long division of 11/13 illustrated below (illustrated by a grade-schooler).

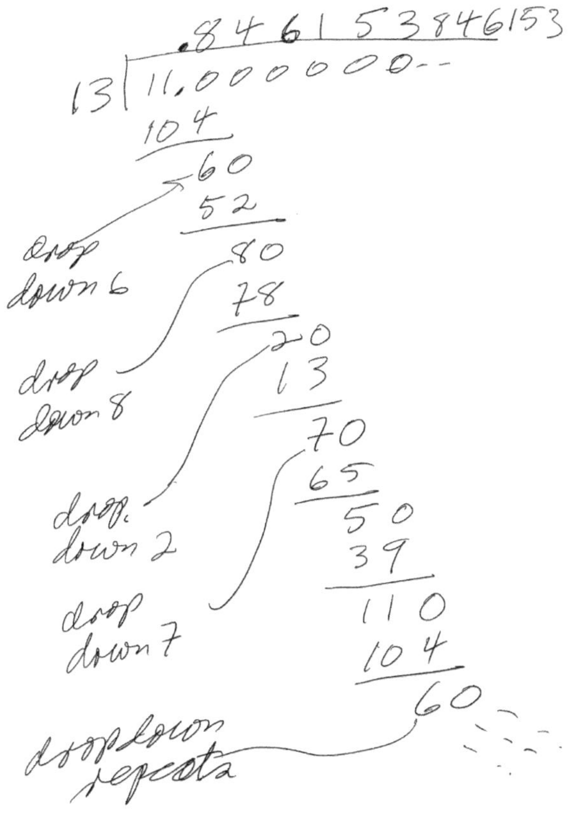

Figure 8.2:

Let d be the **drop down** as illustrated in the above figure. If d = 0, then the decimal representation is truncated by all zeros. If d is not 0, then suppose q is next digit in the remainder in the long division $\frac{x}{y}$. For example, in figure 8.2, when d = 8, then q = 6 and when d = 5, then q = 3. Note that $0 \leq q \leq 9$, and we must have $10d - qy < y$ and therefore $\mathbf{d < y}$. This means that there are only a finite number of possible drop downs, and therefore after a finite number of steps a drop down must equal a previous drop down, in which case the decimal expansion repeats. In the figure, this occurs when d = 11.

8.3.3 the set $\mathbb{Q}_+$ of positive rational numbers is countably infinite

There are obviously an infinite number of positive rational numbers. An infinite *proper* subset of the set of all rational numbers is $\{\frac{1}{2}, \frac{1}{3}, \frac{1}{4}, ...\}$. To see that the set of all rational numbers is countably infinite, we want to find a bijection between $\mathbb{Q}_+$ and $\mathbb{Z}_+$. Picture the rational numbers all being located on the grid, as in the following figure.

In this figure, each vertex corresponds to a rational number. The vertex labeled '2' corresponds to $2 = \frac{2}{1}$, the vertex labeled '3' corresponds to $\frac{1}{2}$, the vertex labeled '9' corresponds to $\frac{1}{4}$, and so forth.

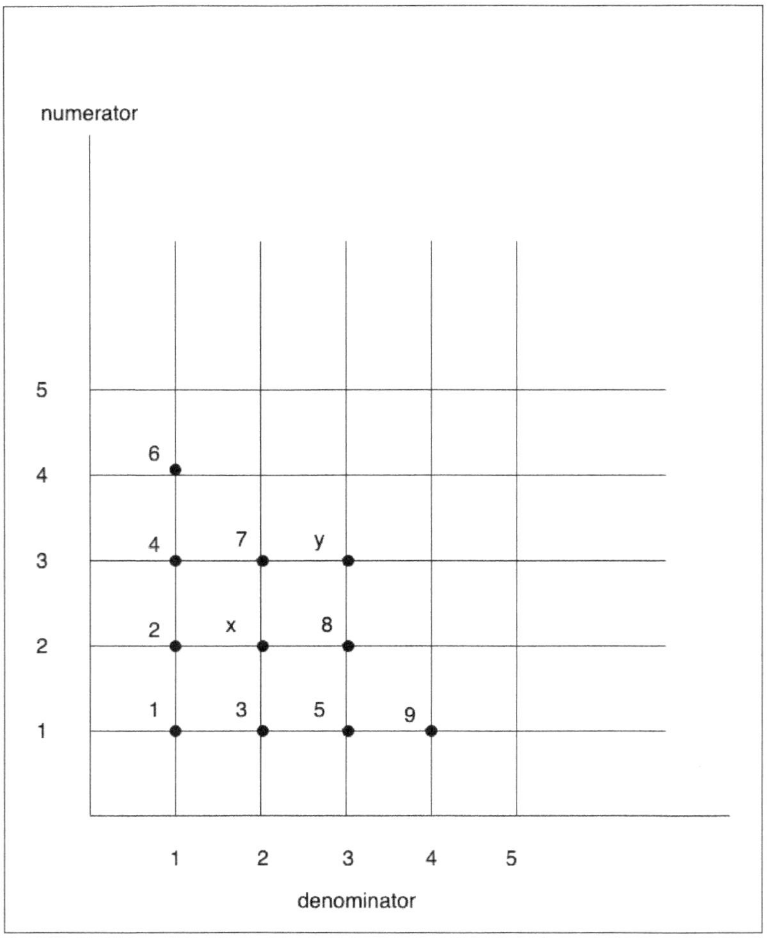

Figure 8.3:

The idea is to count a given rational number only once. That is, only count irreducible fractions. So the vertex labeled 'x' corresponds to $\frac{2}{2} = 1$, which is already accounted for in vertex '1'. The vertex 'y' corresponds to $\frac{3}{3} = 1$, again not counted. A vertex $\frac{4}{8}$ will not be counted as $\frac{1}{2}$ is already included in vertex '3'.

Even though we are not including all vertices, still:

1. There are an infinite number of vertices being counted (all $\frac{1}{n}$, for example), and

2. Every vertex counted is assigned a unique integer label

Therefore every unique rational number is associated with an integer, and every integer is associated with a unique rational number. This is the desired bijection. We can count the rationals, and there are an infinity of them, and so the rationals are countably infinite.

What was just proved is that *there exists* a bijective function

$$\begin{aligned} &F : \mathbb{Z}_+ \to \mathbb{Q}_+ \\ &F(n) = q_n \in \mathbb{Q}_+ \\ &q_n \neq q_{n'}, \quad n \neq n' \end{aligned} \tag{8.14}$$

where q_n is the positive rational number corresponding to the integer n in the bijection.While we have not explicitly constructed the function F, nonetheless we have made an argument to show that such a function exists. This is an **existence proof** and not a **construction proof**, as we have demonstrated the existence of F but have not constructed it explicitly. Many of the most powerful theorems of set theory are existence proofs.

This is an extraordinary result. We have seen that between any two rational numbers there is an infinity of rational numbers, so this bijection seems to be piling infinity on infinity on infinity....

There exists an explicit, constructed bijection from integers to rationals, called the Stern Sequence.

8.3.4 the set $\mathbb{Q}$ of all rational numbers is countably infinite

We have shown that the set of positive rational numbers is countably infinite. We now show that the set of all rational numbers, including 0, is countably infinite, and therefore is isomorphic to $\mathbb{Q}_+$ and hence has cardinality $\aleph_0$. To do this, the bijection

$$G : \mathbb{Z}_+ \to \mathbb{Q} \tag{8.15}$$

is constructed based on the bijection F, equation 8.14 above.

$$\begin{aligned} &G(1) = 0 \\ &G(2) = F(1)(= q_1), \quad G(3) = -F(1)(= -q_1) \\ &G(4) = F(2)(= q_2), \quad G(5) = -F(2)(= -q_2) \\ &\cdots\cdots \\ &G(2n) = F(n)(= q_n), \quad G(2n+1) = -F(n)(= -q_n) \end{aligned} \tag{8.16}$$

where the q_i are the associated rational numbers. Then $\mathbb{Q}^*$ is isomorphic to both since it involves just removing the number 0 from $\mathbb{Q}$.

8.3.5 other countably infinite sets and index sets

A countably infinite set of abstract objects has the form

$$A = \{a_1, a_2, a_3, ...\} \tag{8.17}$$

where the **subscripts** are in 1-1 correspondence with $\mathbb{Z}_+$.

Note that $\mathbb{Z}_+$ itself is a valid index set, so any index set which is isomorphic to this set is also a valid index set. In addition, the set A above can be a countably infinite index set with abstract objects.

8.4 union of countably infinite sets

8.4.1 list-ordered set

In considering the union of countably infinite sets, it can sometimes be useful to think of the elements of a set as being listed in some sort of order. That is, the elements are ordered in some way - smallest to largest, largest to smallest, red to violet and so forth.

I call such a set representation a **list-ordered set**. A list-ordered set is essentially an array, where the set has **slots** and each element is in a slot. Of course a list-ordered set is not a 'real' set, because real sets are unordered. Rather a list-ordered set is sometimes a convenient way of viewing a set in order to simplify some explanation.

So if we define

$$W =< a, b, c, d > \tag{8.18}$$

as an *list-ordered set*, then 'a' is in the first slot, 'b' is in the 2nd slot and so forth. It would be a written representation of the set $\{a, b, c, d\}$. The list-ordered set of integers divisible by 5 would be written

$$< 5, 10, 15, 20, ... > \tag{8.19}$$

where '5' is in the first slot, '10' is in the 2nd slot, etc. The list-ordered set of integers we could notate as

$$< \mathbb{Z}_+ >=< 1, 2, 3, 4, ... > \tag{8.20}$$

> A list-ordered set is exactly the same as an (unordered) set except the elements are viewed in a particular order.

Any list-ordered set **equals** a normal (ie, unordered) set when the two sets have the same elements. **List-ordered sets are used for didactic purposes. Set theory is concerned with unordered sets.**

8.4.2 union of finite set with countably infinite set

Consider the union

$$B = W \cup \mathbb{Z}_+ = \mathbb{Z}_+ \cup W =< a, b, c, d, 1, 2, 3, 4, ... > \tag{8.21}$$

where the union is written as a list-ordered set and W is given by equation 8.18 and where a, b, c, d are not integers. This is a countably infinite set because there is a bijection

$$\begin{aligned} &f : \mathbb{Z}_+ \to B \\ &f(1) = a,\ \ f(2) = b,\ \ f(3) = c,\ \ f(4) = d,\ \ f(n+4) = n,\ \ n \in \mathbb{Z}_+ \end{aligned} \tag{8.22}$$

The point here is that, if the set is list-ordered, **the elements of the finite set are inserted *before* the elements of the infinite set**, and this notion is reflected in the structure of the map f. It would not be meaningful to insert them *after* since the infinite set has no 'end' and therefore 'after' is not meaningful. Of course we could insert them in many different places. Another example of the union might be
$< 1, 2, a, 3, 4, b, c, 5, 6, 7, 8, d, 9, 10, 11, ... >$, but this set is not obviously list-ordered and such notation is not helpful.

8.4.3 Hilbert hotel

An amusing illustration of the nature of a countably infinite set, which uses the above notions, is the Hilbert hotel. The Hilbert hotel is always full and yet there is always room for one more guest, after which it is full again, and so there is room for one more guest, and so forth.

The Hilbert hotel may be modeled by an infinite list-ordered set

$$H_0 =< G_1, G_2, G_3, G_4, ... > \tag{8.23}$$

where guests G_1 (could be one person or a set of one or more people) are in room 1 (slots are called rooms in the Hilbert hotel). Guests G_2 are in room 2, guests G_3 in room 3 and so forth. For every integer n there is a room n, with guests G_n, so the Hilbert hotel has a countably infinite number of rooms and every room is taken. Every room has guests, every guest has a room. The hotel is full.

Suppose new guests B_1 arrive. The hotel manager says that hotel is full but they can always make room for one more party. They move G_1 to room 2, G_2 to room 3, G_3 to room 4, and so forth, and put B_1 into room 1. The guests are now

$$H_1 =< B_1, G_1, G_2, G_3, ... > \tag{8.24}$$

When new guests B_2 arrive, they get room 1, B_1 is moved to room 2, G_1 is moved to room 3, and so forth so the hotel is now

$$H_2 =< B_2, B_1, G_1, G_2, ... > \tag{8.25}$$

It is obvious there is a bijection between $\mathbb{Z}_+$ and any of the arrangements H_k. For H_2

$$F(1) = B_2,\ \ F(2) = B_1,\ \ F(3) = G_1,\ \ F(4) = G_2,\ \ ...., F(n) = G_{n-2}, ...\ n > 2 \tag{8.26}$$

Note that, conceptually, the simplest procedure is that the new guests are put at the head of the line. Of course a new guest could always be put in room N, for any N, but still an infinite number of guests (in rooms N+1, N+2,...) will have to be moved.

The Hilbert hotel may also be modeled by an unordered set of ordered pairs.

$$H_0 = \{(1, G_1), (2, G_2), (3, G_3), ...\} \tag{8.27}$$

where the ordered pair is (roomNumber, guestsId). It might be a useful exercise to reproduce the above reasoning with this type of representation.

8.4.4 union of finite number of countably infinite sets

Theorem 8.10 *The union of a finite number of countably infinite sets is countably infinite. For example, if A, B are countably infinite sets, then the union*

$$R = A \cup B \tag{8.28}$$

R is countably infinite.

Now we are going to use ordinary set notation, but the reader can imagine these as list-ordered sets. Let

$$A = \{a_1, a_2, a_3, a_4, ...\}, \qquad B = \{b_1, b_2, b_3, b_4, ...\} \tag{8.29}$$

We assume all a_i, b_j are distinct, with $a_i \neq b_j$ for any i,j. The resulting set is

$$R = A \cup B = \{a_1, b_1, a_2, b_2, a_3, b_3, ...\} \tag{8.30}$$

R is countably infinite, with bijection $F : \mathbb{Z}_+ \to B$

$$F(n) = \begin{cases} a_{(n+1)/2} & n \ odd \\ b_{n/2} & n \ even \end{cases} \tag{8.31}$$

$F(1) = a_1, F(2) = b_1, F(3) = a_2, F(4) = b_2, ...$

Suppose we have three countably infinite sets

$$A = \{a_1, a_2, a_3, a_4, ...\}, \qquad B = \{b_1, b_2, b_3, b_4, ...\}, \qquad C = \{c_1, c_2, c_3, c_4, ...\} \tag{8.32}$$

The union is

$$R = A \cup B \cup C = \{a_1, b_1, c_1, a_2, b_2, c_2, a_3, b_3, c_3, ...\} \tag{8.33}$$

This is also countably infinite. Exercise: write the bijection explicitly.

Also obviously this generalizes to any finite union of countably infinite sets.

8.4.5 countably infinite union of countably infinite sets

Theorem 8.11

Consider a countably infinite collection of sets

$$S = \{S_1, S_2, S_3, ...\} \tag{8.34}$$

where each S_i is a countably infinite set

$$S_i = \{a_{i,1}, a_{i,2}, a_{i,3}, ...\} \tag{8.35}$$

and where every $a_{i,j}$ is distinct from every $a_{p,q}$ The ***countably infinite union*** *is a set R*

$$R = \bigcup S = \bigcup_{k \in \mathbb{Z}_+} S_k \tag{8.36}$$

where ***R is countably infinite****.*

The analysis is much like the analysis of the set of rationals, where $a_{i,j}$ is analogous to the rational number i/j. This is illustrated in the figure below.

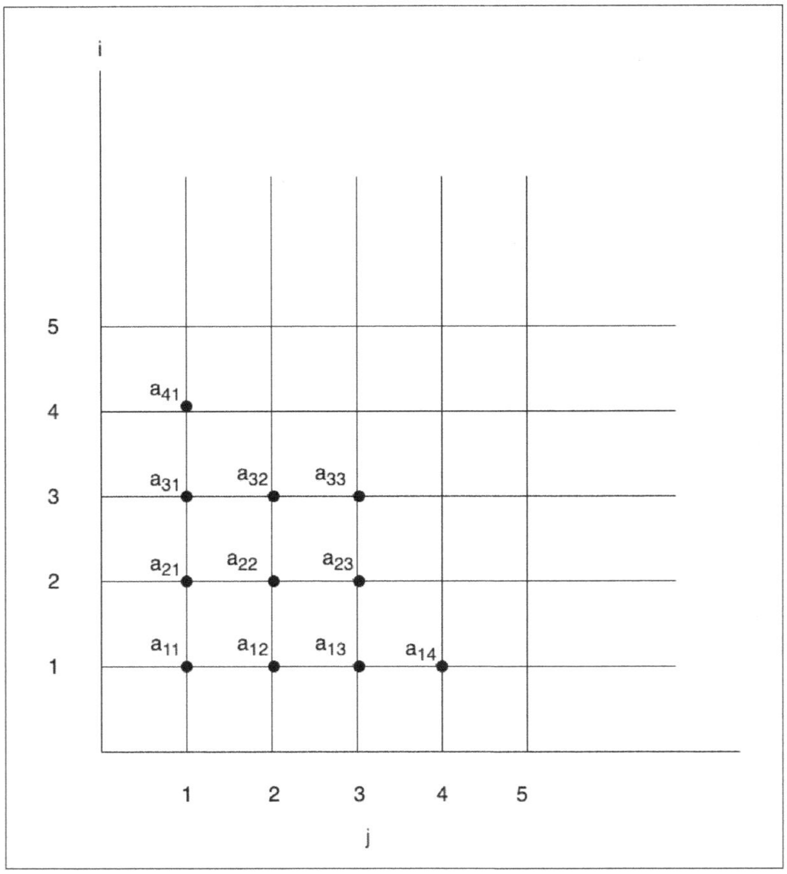

Figure 8.4:

The set R is

$$R = \{a_{1,1}, a_{2,1}, a_{2,2}, a_{3,1}, a_{3,2}, a_{3,3}, a_{4,1}, ...\} \tag{8.37}$$

where the elements of R are listed in the order they are marked in the figure.

8.5 other countably infinite sets of rational numbers

The set of all rational numbers between $1/3$ and $1/2$ is a countably infinite set and is therefore isomorphic to the entire set of positive rational numbers. To properly demonstrate this and other similar, rather paradoxical, isomorphisms, the concept of **function composition** is most useful.

8.5.1 function composition and bijection

We will define **function composition** of *bijective functions* starting with an example. Consider the three sets

$$A = \{a_1, a_2, a_3\}, \quad B = \{b_1, b_2, b_3\}, \quad C = \{c_1, c_2, c_3\} \tag{8.38}$$

all elements distinct. Define functions

$$\begin{aligned} F_{ab} : A \to B, \quad F_{ab}(a_i) = b_i, \; i = 1, 2, 3 \\ F_{bc} : B \to C, \quad F_{bc}(b_i) = c_i, \; i = 1, 2, 3 \end{aligned} \tag{8.39}$$

illustrated in the figure below.

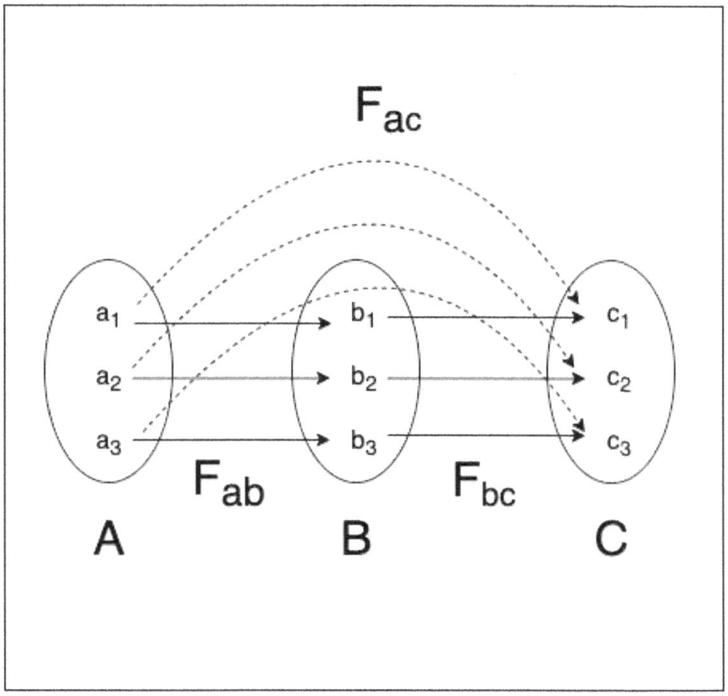

Figure 8.5:

The functions F_{ab}, F_{bc} are bijective.

We may define a function $F_{ac} : A \to C$ **based** on the functions F_{ab}, F_{bc}. Of course we may define a function F_{ac} any way we want, but there is a logical definition which results from the **composition** of the two functions. The idea is to define

$$F_{ac}(a_i) = F_{bc}(F_{ab}(a_i)) = F_{bc}(b_i) = c_i \tag{8.40}$$

which is illustrated in the above figure. This structure is called **function composition** and it is represented abstractly as an expression involving only functions

$$F_{ac} = F_{bc} \cdot F_{ab} \tag{8.41}$$

which has the following meaning:

$$
\begin{aligned}
F_{ac}(a_i) &= (F_{bc} \cdot F_{ab})(a_i) \\
F_{ac}(a_i) &= F_{bc}(F_{ab}(a_i)) = F_{bc}(b_i) = c_i
\end{aligned}
\tag{8.42}
$$

Expression 8.41 is the abstract expression of function composition, while 8.42 is the definition in terms of functions with argument. $\mathbf{F_{ac}}$ **is also bijective.**

Obviously the order of the composition is important. The expression $F_{ab} \cdot F_{bc}$ is not meaningful in the current context because $F_{bc}(a_i)$ is not defined. Function composition is not commutative.

There is function composition of any number of maps. In the case of three maps

$$(f \cdot g \cdot h)(x) = f(g(h(x))) \tag{8.43}$$

Function composition is associative. That is, we may group the functions with parentheses in any way, so long as the order is preserved. If f,g,h are three functions, then

$$
\begin{aligned}
&f \cdot g \cdot h = (f \cdot g) \cdot h = f \cdot (g \cdot h) \\
&((f \cdot g)h(x)) = f((g \cdot h))(x) = f(g(h(x)))
\end{aligned}
\tag{8.44}
$$

Now we express theorem 7.10 in terms of function composition. It is first useful to define an identity map id:

Definition 8.12

Let A be any non-empty set. The ***identity map id*** *is*

$$
\begin{aligned}
&id : A \to A \\
&id(a) = a \ \forall a \in A
\end{aligned}
\tag{8.45}
$$

In other words, **the identity map maps any element of a set to itself** (in the same set). Note

$$id \cdot id = id \Longrightarrow (id \cdot id)(a) = id(id(a)) = id(a) = a \tag{8.46}$$

Now we get back to theorem 7.10. Let $F : P \to Q$ be bijective. Then for $p \in P,\ q \in Q,\ q = F(p)$, we have seen that

$$F(F^{-1}(q)) = q, \ \ F^{-1}(F(p)) = p \tag{8.47}$$

Since this is valid for any p, q where q = F(p), then we can write

$$(F \cdot F^{-1})(q) = q, \quad (F^{-1} \cdot F)(p) = p \tag{8.48}$$

Which, abstractly, is: If F is bijective

$$F \cdot F^{-1} = F^{-1} \cdot F = id \tag{8.49}$$

where id is the identity map.

Finally we may rewrite theorem 7.10 as

Theorem 8.13

A map F is a bijection if and only if

$$F \cdot F^{-1} = F^{-1} \cdot F = id \tag{8.50}$$

Now consider equations 8.38 and 8.39 above. We can see from the above figure that

$$F_{ac}^{-1}(c_i) = a_i \tag{8.51}$$

and based on this it is easy to see that

$$(F_{ab}^{-1} \cdot F_{bc}^{-1})(c_i) = F_{ac}^{-1}(F_{bc}^{-1}(c_i)) = F_{ab}^{-1}(b_i) = a_i$$

and therefore

$$F_{ac}^{-1} = F_{ab}^{-1} \cdot F_{bc}^{-1} \tag{8.52}$$

Note the **reversal of order**. This is a general result.

Theorem 8.14

If two functions $F : X \to Y$, $G : Y \to Z$, are bijective, then the composition

$$H : X \to Z, \quad H = G \cdot F \tag{8.53}$$

is bijective, and

$$\begin{aligned} &H^{-1} : Z \to X \\ &H^{-1} = F^{-1} \cdot G^{-1} \end{aligned} \tag{8.54}$$

is also bijective and is the ***inverse*** *of H.*

To prove this theorem, we can use theorem 8.13 as well as the fact that function composition is associative (equation 8.44). First compose $F \cdot G$ from the left by F^{-1} to get

$$F^{-1} \cdot (F \cdot G) = (F^{-1} \cdot F) \cdot G = id \cdot G = G \tag{8.55}$$

Then compose $F \cdot G$ from the right by G^{-1}

$$(F \cdot G) \cdot G^{-1} = F \cdot (G \cdot G^{-1}) = F \cdot id = F \tag{8.56}$$

Therefore

$$H^{-1} \cdot H = (F^{-1} \cdot G^{-1}) \cdot (G \cdot F) = F^{-1} \cdot id \ \cdot F = id \tag{8.57}$$

which is theorem 8.14. This extends to compositions of more than 2 functions.

For equation 8.44

$$(f \cdot g \cdot h)^{-1} = h^{-1} \cdot g^{-1} \cdot f^{-1} \tag{8.58}$$

Suppose now we have the composition of a bijective function F with a non-bijective function P such that a function Q is defined as

$$Q = F \cdot P$$

Then by the same reasoning as above, we may compose this with F^{-1} from the left to get

$$P = F^{-1} \cdot Q$$

However, $F \neq Q \cdot P^{-1}$ because P^{-1} is a pre-image and not an inverse. In general, $P \cdot P^{-1} \neq id$.

8.5.2 proof of theorem 8.5

Theorem 8.5 states in part that if sets A, B are both countably infinite, they are isomorphic. The assumption is that there exist bijections $F : \mathbb{Z}_+ \to A$ and $G : \mathbb{Z}_+ \to B$. Consider

$$\begin{aligned} &(F \cdot G^{-1}) : B \to A \\ &(F \cdot G^{-1})(b_i) = F(G^{-1}(b_i)) = F(n_i) = a_j \end{aligned} \tag{8.59}$$

where $b_i \in B$, $a_j \in A$ and $n_i \in \mathbb{Z}_+$. Since G^{-1} is bijective, then the composition $F \cdot G^{-1}$ is bijective, from theorem 8.14. The second of equations 8.59 is a bijection between A and B, so the sets are isomorphic.

The other part of theorem 8.5 is this: If sets A, B are isomorphic and A is countably infinite, then B is countably infinite. The assumption is that there exist bijections
$F : A \to B, \ G : \mathbb{Z}_+ \to A$. Then

$$(F \cdot G) : \mathbb{Z}_+ \to B \tag{8.60}$$

Since, from theorem 8.14, $F \cdot G$ is a bijection, then B is countably infinite.

8.5.3 proof of theorem 7.12

This theorem states that if $A \cong B$ and $B \cong C$, then $A \cong C$. This has already been proven (cf figure 8.5.) This is a simple composition. By assumption, there exists bijections $F : A \to B$, $G : B \to C$. Then $H = G \cdot F$ is bijective and $H : A \to C$, so $A \cong C$.

8.5.4 other sets of rational numbers

First a definition.

Definition 8.15 ***notation***

*Let a, b be rational numbers, with $a < b$. The set ((a,b)) is the set of all **rational** numbers r between a, b exclusive: ie*

$$a < r < b \tag{8.61}$$

*The set ((a,b)) is called an **open** interval of rational numbers. The set can be defined as*

$$((a, b)) = \{r | r \in \mathbb{Q} \,\&\, a < r < b\}, \quad a, b \text{ rational} \tag{8.62}$$

In this notation, $\mathbb{Q}_+ = ((0, \infty))$ and $\mathbb{Q} = ((-\infty, \infty))$

*We can define [[a,b]] as a **closed** interval of rational numbers*

$$[[a, b]] = \{r | r \in \mathbb{Q} \ \&\ a \leq r \leq b\}, \quad a, b \text{ rational} \tag{8.63}$$

We have already shown that $\mathbb{Q}$, $\mathbb{Q}_+$, $\mathbb{Q}^*$ are isomorphic, and that $\mathbb{Q}_+$ is countably infinite. Therefore, from theorem 8.5, these sets are each countably infinite.We now show

Theorem 8.16

The set ((0,1)), the rational numbers between 0 and 1 (exclusive), is countably infinite. Therefore this set is isomorphic to $\mathbb{Q}$ and to $\mathbb{Q}_+$.

We note that

$$((0, 1)) \subset \mathbb{Q}_+ \subset \mathbb{Q} \tag{8.64}$$

The above theorem states that these three sets have the same cardinality - ie, $\aleph_0$. Here again is the paradoxical behavior of infinite sets. There should be "fewer" rational numbers between 0 and 1 than there are all rational numbers, yet the sets are in 1-1 correspondence. (The reason, of course, is that the word "fewer" in this context is meaningless.)

To prove this theorem, the following is a bijection between ((0,1)) and $\mathbb{Q}_+$.

$$\begin{aligned} &F : ((0,1)) \rightarrow ((0,\infty)) \\ &q = F(x) = \frac{x}{1-x}, \quad x \in ((0,1)), \quad q \in \mathbb{Q}_+ \end{aligned} \tag{8.65}$$

Theorem 8.17

For any rational numbers a, b, a < b, the set ((a,b)), the rational numbers between a and b (exclusive), is countably infinite.

To prove this, we show that ((a,b)) is isomorphic to ((0,1)). The bijection F between the sets is

$$x = F(q) = \frac{q-a}{b-a}, \quad x \in ((0,1)), \quad q \in ((a,b)) \tag{8.66}$$

In other words, **any two open intervals of *rationals* ((a,b)), ((c,d)), a < b, c < d, are isomorphic and countably infinite**.

Theorem 8.18

The ***closed interval*** *of rationals* $[[a,b]]$ *is countably infinite and therefore is isomorphic to any other open or closed interval of rationals.*

Exercise: prove this using equation 8.66.

Exercise: Make up similar theorems about **half-open intervals** such as $((a,b]] = \{r | r \in \mathbb{Q} \,\&\, a < r \leq b\}, a < b,\ a, b$ finite and rational.

8.6 another countably infinite set

This section could be considered a digression.

> **Theorem 8.19**
>
> *Let S be any countably infinite set. Then the set U of all* ***finite subsets*** *of S is countably infinite.*

To show this, it suffices to prove it for all finite subsets of $S = \mathbb{Z}_+$. Define sets

$$\begin{aligned} S_1 &= \{\{1\}, \{2\}, \{3\}, ...\} \\ S_2 &= \{\{1,2\}, \{1,3\}, \{1,4\}, ..., \{2,3\}, \{2,4\}, \{2,5\}, ... \\ S_3 &= \{\{1,2,3\}, \{1,2,4\}, ...\} \\ &etc \end{aligned} \tag{8.67}$$

Then

$$U = \bigcup_{i \in \mathbb{Z}_+} S_i = \{\{1\}, \{2\}, ..., \{1,2\}, \{1,3\}, ...\{1,2,3\}, \{1,2,4\}, ...\} \tag{8.68}$$

is the set of all finite subsets of $\mathbb{Z}_+$. Each set S_i is countably infinite. This can be proved using induction. Obviously S_1 is countably infinite. Then to show S_2 is countably infinite we use the abstract array $A_{1,2} \rightarrow \{1,2\}, A_{1,3} \rightarrow \{1,3\}$, and so forth. Then from section 8.4.5 the set of $A_{i,j}$ is countably infinite, and since this set is isomorphic to S_2, then S_2 is countably infinite.

For S_3, we have the bijection $F : S_2 \rightarrow \mathbb{Z}_+$ from above. Assume for purposes of illustration that $F(1) \rightarrow \{1,2\}, F(2) \rightarrow \{1,3\}$, and so forth. .Therefore S_3 is isomorphic to the abstract array $B_{F(1),3}, B_{F(1),4}, ..., B_{F(2),4}, B_{F(2),5}, ...$ and so forth. Then from theorem 8.4.5, S_3 is countably infinite.

By induction we see that all S_i are countably infinite. (It is a good exercise to tighten up this argument.)

Therefore U is the countably infinite union of countably infinite sets, which, from theorem 8.4.5, is countably infinite.

We can see that **countably infinite is a very big category**. While $\mathbb{Z}_+$ is the prototype of a countably infinite set (it *defines* countably infinite), still we can create a bijection from $\mathbb{Z}_+$ to a huge number of sets, all of which have the same cardinality.

9 real numbers. Uncountable sets.

The **set of real numbers** is the set of all rational and irrational numbers. The irrational numbers are defined below. A **real number** is either a rational or irrational number.

The set of real numbers is also called a **continuum**.

First it will be useful to review the notation for numbers in an arbitrary base.

9.1 numbers in any base

Definition 9.1

Any real number x ***represented in base B*** *(B ≥ 2 an integer) is of the form*

$$x = n_N n_{N-1} ... n_1 n_0 \ . \ x_1 x_2 x_3 .....|_B$$

where all n_i, x_i are digits between 0 and B-1 and $N \geq 0$ is finite. If $N = 0$ and $n_0 = 0$, the number is between 0 and 1. The $|_B$ indicates that the number is written to base B. If the base is understood, this may be left out.

The ***decimal*** *(base 10) representation of this number is*

$$n = \sum_{i=0}^{N} n_i B^i + \sum_{k=1}^{\infty} \frac{x_k}{B^k} \tag{9.1}$$

This is a generalization of definition 8.3.2 where we are not specifically considering rational numbers.

If B = 10, this is the **decimal** system.

If B = 2, this is the **binary** number system. The number $101|_2$ in the decimal system is $101|_2 = 1 * 2^2 + 0 * 2^1 + 1 * 2^0 = 4 + 1 = 5$. The number $0.11001|_2$ in the decimal system is $0.11001|_2 = \frac{1}{2} + \frac{1}{2^2} + \frac{0}{2^3} + \frac{0}{2^4} + \frac{1}{2^5} = 0.5 + 0.25 + 0.03125 = 0.78125$.

A **rational** number is one where there is a repeating termination, regardless of base. So in binary, 0.10011010011001100110011001..... is a rational number with the repeat 1001. Also in binary the number 0.1001101011111111111......... is a rational number and is equal to 0.10011011. The rationale for both these cases are identical to that given in the definition 8.3.2 except using 2 instead of 10 in the denominators and 1 instead of 9.

9.2 irrational numbers vs rational numbers. Real numbers.

Definition 9.2

An ***irrational number*** *is a number which cannot be written as the quotient of two integers. That is, an irrational number is any number which is not rational.*

$\sqrt{2}$ is an irrational number.

To prove this, suppose $\sqrt{2}$ were rational. Then we could write $\sqrt{2} = \frac{a}{b}$, where a, b are positive integers and $a > b$ (since $\sqrt{2} > 1$). We assume that this fraction is fully reduced. That is, we assume 'a' and 'b' do not have a common divisor.

Then $2 = \frac{a^2}{b^2}$ or $a^2 = 2b^2$. This means that a^2 is even, which means that 'a' must be even, so a = 2p, where p is an integer. Then $4p^2 = 2b^2$ which means $b^2 = 2p^2$. Therefore b^2 is even, which means 'b' is even. Since both 'a' and 'b' are even, then 'a' and 'b' have a common divisor (2) which is a contradiction, and therefore $\sqrt{2}$ is irrational.

If p is a prime number, then $\sqrt{p}$ is an irrational number. This is proved the same way as for $\sqrt{2}$, as follows. If a,b are integers and $a^2 = pb^2$, where p prime, then a=pw, where w is an integer. (Suppose this were not true. Since any number is a product of primes, then $a = q_1q_2q_3...$, all q_i prime. Since p is not part of this product, then p cannot be a factor in a^2, a contradiction.) The proof that $\sqrt{p}$ is irrational then follows.

It has been proven that both π and ℓ (exponential) are irrational numbers, although the proof is beyond the scope of this book. The subject of irrational numbers - what is irrational and what is not irrational - is not settled. Any number can be written in the form

$$n_1n_2n_3....n_N.x_1x_2x_3x_4..... \qquad (9.2)$$

a finite sequent of digits n_i followed by a decimal point followed by a countably infinite sequence of digits x_i. In other words, definition 9.1 implies that

Any number can be written as a countably infinite sequence of digits. This means that any number is isomorphic to a countably infinite *set* of digits written $\{(1, x_1), (2, x_2), ...\}$

It was shown that if x is a number with a repeating termination, then x is rational. The contrapositive is that **if a number x is not rational, then it does not have a repeating termination**. Therefore

Theorem 9.3

A number x is irrational iff it does **not** *have a repeating teremination.*

Now consider only numbers between 0 and 1. For any **rational** number r, 0 < r < 1, we can write (definition 8.3.2)

$$r = 0.z_1z_2z_3.... = 0.y_1y_2y_3...y_kx_1x_2...x_jx_1x_2...x_jx_1x_2...x_j..... \tag{9.3}$$

any finite sequence of digits y_i followed by a countably infinite repeating sequence of digits $x_1x_2...x_j$. It is possible that k=0 (so there are no y_i). It is also possible that j=1, as in 0.33333....

On the other hand, an irrational number t (t < 1) has the form

$$t = 0.y_1y_2y_3.... \tag{9.4}$$

a countably infinite sequence of digits y_i where there is **no repeating termination**. For example, the toy number 0.1234567891011121314151617.... is an irrational number between 0 and 1, as is 0.010203040506070809010011012.... .

If t is irrational and r is rational, then the product rt is irrational. Suppose the product were rational. Then rt=a/b, for a,b integers. Therefore t=a/(br), which is rational, a contradiction.

If t is irrational and r is rational, then t+r is irrational. For if not, then t+r=a/b so t = (a/b-r)=(a-br)/b, which is rational, a contradiction.

However, if u, v are irrational, then uv or u+v may or may not be irrational. For example, $u = 1 + \sqrt{2}$, $v = 1 - \sqrt{2}$, then u+v=2, rational.

Similarly note that $\sqrt{18} = 3\sqrt{2}$ is irrational but $\sqrt{2}\sqrt{18} = \sqrt{36} = 6$ is rational.

There are several sets of numbers that are conventionally defined. The **algebraic real numbers** are all the real solutions to all polynomial equations of the form

$$\sum_{n=0}^{N} a_n x^n = 0 \tag{9.5}$$

for all $N \in \mathbb{Z}_+$ and a_n are integers. For example, $x^2 - 2 = 0$ has solutions $x = \pm\sqrt{2}$. Of course, many solutions can be rational, such as the solution to $x - 1 = 0$. Therefore the (real) algebraic numbers include all the rational numbers (which includes integers) and some irrational numbers. The set of algebraic numbers must include at least a countably infinite number of irrational numbers since all $\sqrt{p}$, p prime, are irrational and there are a countably infinite number of primes. (There are an infinite number of primes, the set of primes is a proper subset of the set of integers, and therefore the set of primes is countably infinite. See theorem 8.6.)

The set of algebraic numbers is countably infinite. Consider the polynomial equation 9.5 for given (finite) N. For any given set of a_n, there are a maximum N real solutions. The number of different sets of a_n, for given N, is countably infinite, since all a_n are integers. Thus for given N there are at most a countably infinite number of real solutions. The set of real algebraic numbers is then a countably infinite union of the above countably infinite sets (one for each N) which, as has been shown, is countably infinite. The **transcendental numbers** are all the *irrational* numbers which are not algebraic numbers. So π and ℓ are transcendental numbers. We will see that the cardinality of the set of transcendental numbers is a different cardinality than the cardinality of a countably infinite set. (Later we show that it is 'larger'.)

The **set of real numbers is the union of the set of algebraic numbers and the set of transcendental numbers**.

The **real numbers** are all the numbers familiar from usual arithmetic. These are all numbers between $-\infty$ and ∞. The symbol for all real numbers is $\mathbb{R}$. The symbol for all positive real numbers is $\mathbb{R}_+$ (or $\mathbb{R}^+$. We will use the former.) The symbol for all real numbers excluding 0 is $\mathbb{R}^*$.

The real numbers have several **squeeze properties**. Between any two reals there is another real. In fact, between any two irrationals is another irrational and a rational. Between any two rationals is an irrational (as well as a rational). It is a nice exercise to prove this. (Say $x_1 < x_2$ are irrational. If $q = \frac{x_1+x_2}{2}$ is rational, what is $\frac{q+x_2}{2}$?)

Sometimes the set of **reals** (reals = real numbers) is called the **real line** and is pictured as an infinitely long line. Every point on that line corresponds to a real number. (See the discussion later.)

9.3 the cardinality of the set of real numbers between 0 and 1

We will now show that the set of real numbers is not isomorphic to the set of integers. That is, the set of real numbers is not countably infinite and is said to be **uncountable**.

First consider the set of all real numbers in (0,1), and **assume** this set is countably infinite. Then we ought to be able to write that set of real numbers in the decimal representation as follows (**the initial 0 and decimal point are suppressed for clarity**):

$$\begin{aligned} w_1 &= a_{1,1}a_{1,2}a_{1,3}\ldots\ldots \\ w_2 &= a_{2,1}a_{2,2}a_{2,3}\ldots\ldots \\ w_3 &= a_{3,1}a_{3,2}a_{3,3}\ldots\ldots \\ &\ldots\ldots\ldots \\ &\ldots\ldots\ldots \end{aligned} \tag{9.6}$$

where $a_{i,j}$ is a digit, one of 0,1,2,...,9. Since the assumption is that the above set is countable, we 'count" the elements using a countably infinite number of variables $w_1, w_2, ...$ (w_1 is really $0.a_{1,1}a_{1,2}a_{1,3}......$ for example).

It is possible that two numbers have the same sequence for the first finite set of digits, but the final sequence of one of the numbers is 00000....and the final sequence of the other is 99999.... For example, 0.423000000000.... and 0.422999999..... These numbers are the same number (definition 8.3.2) and the above set, equation 9.6 is assumed to only contain the terminating sequence of all zeros (as appropriate, obviously). That is, in our example, the number 0.42300000000 is included in the above list but .42299999 is not included.

Definition 9.4

*Consider two numbers $x = 0.x_1x_2x_3.....$, $y = 0.y_1y_2y_3......$ The two numbers are **equal** when, for every i, $x_i = y_i$. Therefore, if the two numbers differ in the digits of even one corresponding subscript, they are **not equal**.*

We can now prove that the putative set of 'all' reals, equation 9.6, cannot contain all the real numbers between 0 and 1. We prove this by constructing a number which is not included in that set. In other words, for the **countably infinite** set $W = \{w_1, w_2, w_3, ...\}$, we will show that no map $F : W \rightarrow \mathbb{R}$ can be surjective and therefore no map can be bijective and therefore the sets W and $\mathbb{R}$ are not isomorphic.

Let q be the number we construct. Say $q = 0.q_1q_2q_3.....$ We define each digit q_k as follows.

Consider the first number $0.a_{1,1}a_{1,2}a_{1,3}......$ If $a_{1,1}$ is equal to 1 set q_1 equal to 2. Otherwise set q_1 to 1.Therefore $q_1 \neq a_{1,1}$.

Do similarly for q_2. If $a_{2,1}$ equals 1 set $q_2 = 2$ otherwise set $q_2 = 1$. Therefore $q_2 \neq a_{2,2}$.

Proceeding in this way for all the numbers of equation 9.6, the number q **might** look like 11222121122122112221111122...., or some other string of 1s and 2s. But q differs from each number in equation 9.6 in at least one digit. Thus q differs from number w_1 in the first digit, it differs from w_2 in the second digit, it differs from w_3 in the third digit, and so forth.

Therefore, the number q, 0 < q < 1, is not included in the set of numbers equation 9.6. Since this countably infinite set was assumed to contain **all** reals between 0 and 1, that assumption is not correct. Therefore **the set of real numbers between 0 and 1 is not countably infinite**. Its cardinality is said to be **uncountable**.

The cardinality of the set (0,1) of reals is represented by the symbol $\aleph_1$, different from $\aleph_0$.

9.4 cardinality of intervals of real numbers

We define intervals of real numbers.

Definition 9.5

Let a,b be real numbers, with $a < b$.

[a,b] is a set which includes all real numbers t where $a \le t \le b$. This is a ***closed interval****.*

(a,b) is the set which includes all real numbers t where $a < t < b$. This is an ***open interval****.*

(a,b] is the set which includes all real numbers t where $a < t \le b$. This is a ***half-open interval****.*

[a,b) is the set which includes all real numbers t where $a \le t < b$. This is a ***half-open interval****.*

Note that notation such as $[-\infty, \infty]$, $(a, \infty]$, $[-\infty, b]$ is meaningless. ∞ is a symbol and not a number and so cannot be **in** any sets of reals.However, we can *conveniently* notate

$$\mathbb{R} = (-\infty, \infty), \quad \mathbb{R}_+ = (0, \infty) \tag{9.7}$$

because the notation $(-\infty, \infty)$ is defined to indicate that every finite real number is in this set. The set given by the second expression of equation 9.7 contains all positive real numbers.

Note that $\mathbb{R}^* = (-\infty, 0) \cup (0, \infty)$ is **not an interval**. It is the union of two disjoint intervals.

The following intervals should be self-explanatory:

$$(-\infty, a), \; (-\infty, a], \quad (b, \infty), \; [b, \infty) \tag{9.8}$$

Intervals are sets of real numbers and therefore we can define set membership and subsets and take unions and intersections. For example

$$\begin{aligned}
&[1, 2] = [1, 1.5] \cup [1.5, 2] \\
&[a, b] = [a - x, b + y] \cap [a, b], \quad x, y \ge 0 \\
&(-1, 0) \cap (0, 1) = \emptyset \\
&\mathbb{R}^* = (-\infty, 0) \cup (0, \infty) \\
&(2, 5] \subset [2, 6), \quad (2, 5] \subseteq (2, 5] \\
&0.25 \in (0, 1), \quad 0 \notin (0, 1), \quad 1 \notin (0, 1)
\end{aligned} \tag{9.9}$$

In the third and fourth lines above, 0 is not in the set. The fifth line, 2nd expression, means the set is a subset of itself.

Note that $(-1,0) \cup (0,1)$ is not an interval. The two intervals being unioned are disjoint. Note also that $[1,3] \cup (5,6)$ is not an interval.

In other words,

> The union of any two intervals is not necessarily an interval.
>
> The intersection of any two intervals is either an interval, a number or the empty set.

An exercise is to prove the above.

What numbers are in the interval (0,1)? We can divide the real numbers into two categories: rational numbers and irrational numbers. Since the set of rationals in (0,1) is countable, then if the set of irrationals in (0,1) were countable, the interval (0,1) would be countably infinite. But we just proved that such is not the case.Therefore the set of irrational numbers in (0,1) is uncountable.

We can divide the irrationals into two categories: The algebraic irrationals and the transcendentals. The algebraic irrationals are countably infinite, as discussed above. If the transcendentals were countably infinite, then all the irrationals would be countably infinite, which is not the case.

> **The transcendental irrationals are uncountable**. The set of all the other numbers (including the set of algebraic irrationals) is countably infinite.

The transcendentals make up the vast majority of numbers in (0,1) and, in fact, in any set of intervals of the real line. Poetically, the rationals are pebbles in the continuum 'river' of irrationals. While there is an infinity of rationals in (0,1) there is a different infinity of irrationals.

One interesting difference between the rationals and the reals is in the intervals. Let $[[0,\sqrt{2}]]$ be the set of *rationals* between 0 and $\sqrt{2}$ inclusive (we are somewhat abusing the notation of section 8.5.4). $[0,\sqrt{2}]$ is the set of *reals* in the same interval. The set $[[0,\sqrt{2}]]$ is in fact a **half-open interval** of rational numbers, because it includes all rationals but no rational number equals $\sqrt{2}$. On the other hand, $[0,\sqrt{2}]$ is a **closed** interval (of real numbers). This difference turns out to be of some significance.

9.5 any interval of real numbers is uncountable

This is the same concept as for rationals, but now the bijections can be a little more tricky. We start with a simple case. We show

$$(0,1) \cong \mathbb{R}_+ \tag{9.10}$$

In other words, (0,1) is isomorphic to $\mathbb{R}_+$ and hence $\mathbb{R}_+$ is uncountable. We might not have expected this since (0,1) is a bounded interval while $\mathbb{R}_+$ is unbounded. The bijection is:

$$\begin{aligned} &G : (0,1) \to \mathbb{R}_+ \\ &G(x) = \frac{x}{1-x}, \quad 0 < x < 1 \end{aligned} \tag{9.11}$$

Note that $\frac{x}{1-x}$ is rational if x is rational and irrational if x is irrational. (Suppose x is irrational but $\frac{x}{1-x}$ were rational. Then $\frac{x}{1-x} = \frac{a}{b}$ and therefore $x = \frac{a}{a+b}$, which is rational and a contradiction.)

Here is one that is a little trickier.

$$\mathbb{R} \cong (0,1) \tag{9.12}$$

In other words, the set of all reals (the whole real line) is in 1-1 correspondence with the open interval (0,1). The bijection between the sets is:

$$\begin{aligned} &F : (0,1) \to \mathbb{R} \\ &F(x) = \begin{cases} \frac{x-0.5}{x}, & 0 < x < 0.5 \\ \frac{x-0.5}{1-x}, & 0.5 < x < 1 \\ 0 & x = 0.5 \end{cases} \end{aligned} \tag{9.13}$$

That is, (0,0.5) is mapped to negative reals, (0.5,1) is mapped to positive reals, and 0.5 is mapped to 0.

Both F, equation 9.13, and G , equation 9.11, are bijections and therefore

$$F \cdot G^{-1} : \mathbb{R}_+ \to \mathbb{R} \tag{9.14}$$

is a bijection (Exercise: prove this), therefore

$$\mathbb{R}_+ \cong \mathbb{R} \cong (0,1) \tag{9.15}$$

so that $\mathbb{R}_+$ and $\mathbb{R}$ are isomorphic and so both are uncountable (cardinality $\aleph_1$).

Next we show

$$[0,1) \cong (0,1) \tag{9.16}$$

the half-open interval is isomorphic to the open interval. This illustrates how the difference between the rationals and irrationals plays in the intervals of real numbers. Let [[0,1)) be the set of rationals r, $0 \leq r < 1$, and ((0,1)) be the open interval of rationals (see definition

8.15). As discussed in section 8, a bijection H exists because the two sets are countably infinite:

$$H : ((0,1)) \to [[0,1)) \tag{9.17}$$

Then a bijection K between intervals (0,1) and [0,1) of reals is

$$\begin{aligned} &K : (0,1) \to [0,1) \\ &K(x) = \begin{cases} H(x), & x \; rational \\ x, & x \; irrational \end{cases} \end{aligned} \tag{9.18}$$

An equivalent argument yields the isomorphisms

$$(0,1) \cong [0,1) \cong (0,1] \cong [0,1] \tag{9.19}$$

Just the map H, equation 9.17, changes in detail. All these sets are isomorphic and therefore all of cardinality $\aleph_1$.

Finally, for completeness, we show that

$$(0,1) \cong (a,b) \tag{9.20}$$

where a < b real.The following is the desired bijection.

$$\begin{aligned} &M : (0,1) \to (a,b) \\ &M(x) = x(b-a) + a. \end{aligned} \tag{9.21}$$

It follows that any interval (a,b) is isomorphic to any of the intervals, equation 9.19. The same argument shows (a,b] is isomorphic to (0,1], and hence $(a,b] \cong (a,b)$, and so forth. Therefore any of the intervals between a-b is isomorphic to any of the intervals between c-d, and we have the following result:

Theorem 9.6

Any of the intervals of definition 9.5 and of equations 9.7 and 9.8 are isomorphic to any other such interval.

We conclude: ***any interval of real numbers is uncountable.***

All possible cases for intervals have not been covered, but I think the coverage is sufficient.

9.6 discussion

The isomorphisms of intervals of reals is equivalent to **scaling** intervals of the real line. Consider the figure below.

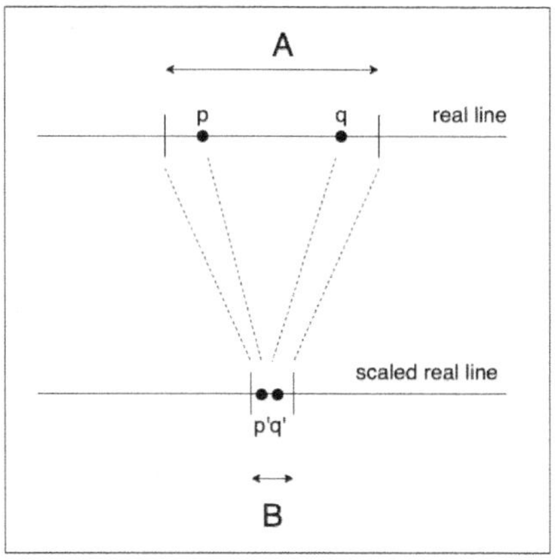

Figure 9.1:

The interval A is shrunk (scaled) down to interval B. Two points of A - say p, q - are shifted continuously to the points p', q' of B. The points are distinguishable no matter the amount of scaling. Further, if p and q are rational and the interval A is scaled by an irrational number, the points p', q' could become irrational, or one could be rational and the other irrational, or both rational.

This property becomes *subjectively uncomfortable* if we imagine interval A as very large (say 10 light years) and interval B as tiny (10^{-16} cm). As one interval is scaled to the other, not a single point is lost or gained. The points of the two intervals are in 1-1 correspondence (the intervals are uncountable and isomorphic). Every point on the upper line can be traced to a corresponding point on the scaled line.This of course is a crucial feature of calculus and many other areas of mathematics.

This is equivalent to the assumption of Euclidean geometry that a point has no dimension. A point has no width. Nonetheless a continuum of points does have a width (the width of the interval [0,1] is 1).

Is it sensible that two distinct point physical objects (such as electrons appear to be) can occupy any two arbitrary points in space, no matter how close the points, with non-vanishing probability? Since, mathematically, the isomorphisms presented in this section are all logically consistent, it would appear that the applicability to a real world (such as the one we live in) would be a problem in physics.

10 finite unions of uncountable sets

Theorem 10.1

The finite union of uncountable sets is uncountable.

For example, the set

$$S = (2,3) \cup (5,6) \tag{10.1}$$

is uncountable. To show this, set up a bijection between S and, for example, (0,1) by mapping (2,3) into (0,.5) and mapping (5,6) into [0.5,1). We will not belabor this here.

10.1 The Cantor set

Of all the subsets of the real line, the Cantor set is probably the most fertile source of examples and counterexamples in topology. Fred H. Croom

So far we have discussed uncountable sets in the context of intervals of the real line as well as the whole real line. However, **not all uncountable sets are intervals or unions of intervals**, and the Cantor set is one such set.

Consider the interval [0, 1] and the **ternary** (base 3) representation of a number n=$0.x_1x_2x_3....|_3 \in [0,1]$, where $x_i \in \{0,1,2\}$. In the decimal representation the number n is

$$n = \sum_{i=1}^{\infty} \frac{x_i}{3^i} \tag{10.2}$$

We assume no termination of all 2's. Now create a **proper** subset C

$$C \subset [0,1] \tag{10.3}$$

where, for every number in C, $x_i \in \{0,2\}$. That is, there are no 1's in the ternary representation of any number in set C. The decimal representation of the numbers in C is still given by equation 10.2.

First, C is uncountable because the real numbers in C can be put in 1-1 correspondence with the binary representation of a number (not the same number, and no termination of all 1's for binary). Obviously C is a proper subset of [0, 1] because some numbers in [0, 1] are not in C but every number in C is in [0, 1]. So we have C as a proper subset of [0, 1] and yet uncountable.

We now consider any interval $(a,b) \subset [0,1]$, 0 < a < b < 1. Clearly

$$(a,b) \notin C \tag{10.4}$$

because there are numbers in any interval (a,b), a $<$ b, which include 1 in the ternary representation and such numbers are not in C. The same holds for closed and other intervals.

> Even though C is uncountable, **C contains no intervals**. The set C is the **Cantor set**, and is sometimes referred to as **dust**. It is a fractal.

The set C has many other paradoxical properties, which are outside the scope of this book.

10.2 uncountable index sets

Briefly, an index set may be uncountable. The sets $\mathbb{R}_+$, $(0, 1)$ and many others may be index sets. Then there can exist some set S whose elements are subscripted and the subscripts refer to the contents of an uncountable index set. The set S may be finite, countably infinite or uncountable, depending on how many elements it actually has.

As a simple example, suppose $I = \mathbb{R}_+$ is an index set and define g_x as some function of x (for example, a sine function) where $x \in I$. Then a valid finite set S is $\{g_{1.2}, g_{2.5}, g_{3.8}\}$. Another possibility is the uncountable set $S = \{g_x\}|x \in (0, 1)$.

11 the algebra of sets

In this section the nuts and bolts calculations of set theory are introduced. The material of this section is used anywhere that sets are important, so it is important to at least be aware of this material. One could quickly read this section on a first pass through the text, but eventually the material should be absorbed.

11.1 Venn diagrams

In section 4.2 we discussed what we called "diagrams of sets". A **Venn diagram** is another type of diagram providing **visualization of sets**. While the diagrams of section 4.2 represent specific sets, by listing the elements of the sets, a Venn diagram represents relationships among generic sets.

For example, the set A, whatever it is, would be represented by one of the diagrams below.

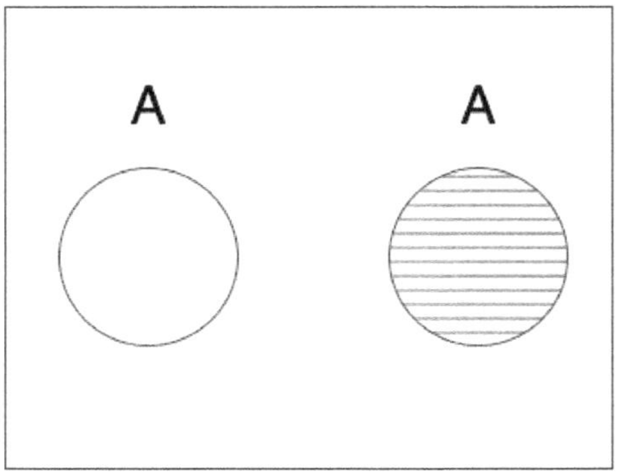

Figure 11.1:

The diagram on the left, with no shading, might be used to show general relationships among sets, while the shading of the diagram on the right represents the contents of set A.

The figure below

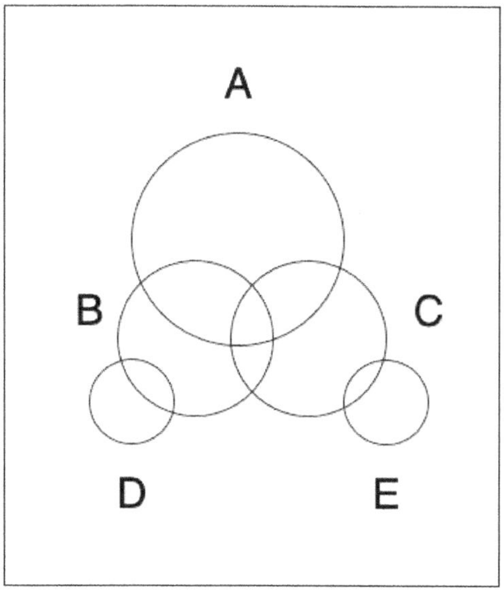

Figure 11.2:

shows a relationship among five sets. The diagram is drawn such that sets A, B and C pairwise intersect, while B and D intersect and C and E intersect. According to this figure, $E \cap A = \emptyset$, $D \cap A = \emptyset$ and so forth.

Adding lines provides a visualization of the result of a particular expression. The shaded area (area with lines) of the figure below

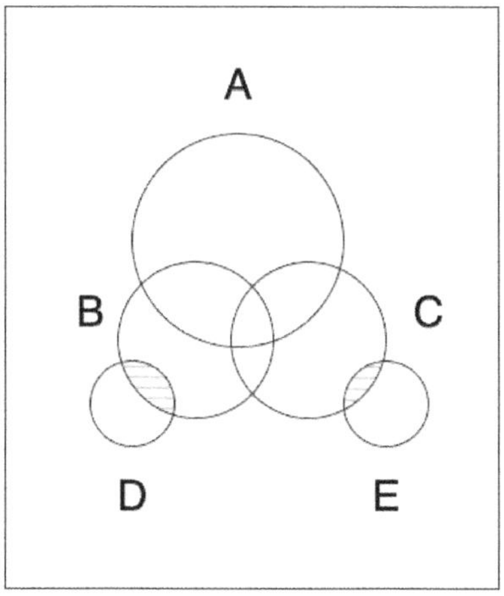

Figure 11.3:

represents all the points of the expression

$$(B \cap D) \cup (C \cap E) \tag{11.1}$$

On the other hand, the shaded area of the figure below.

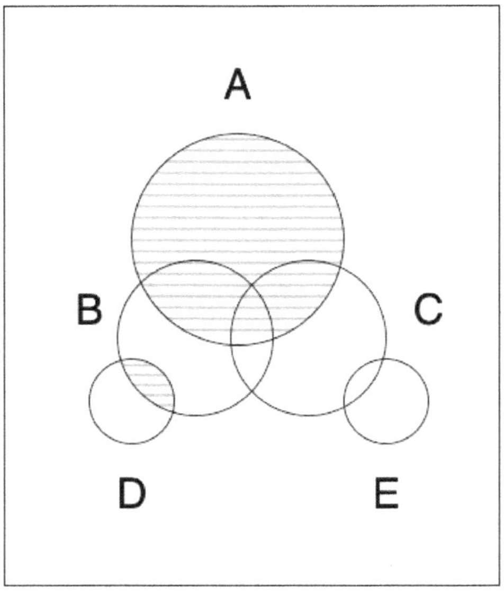

Figure 11.4:

represents all the points of the expression

$$A \cup (B \cap D) \tag{11.2}$$

11.2 intersections, unions and Venn diagrams

11.2.1 Venn diagram of $\mathbf{A} \cap \mathbf{B}$

Venn diagrams represent the contents of sets with slanted or horizontal lines. Let us agree to use the following terminology and abbreviations:

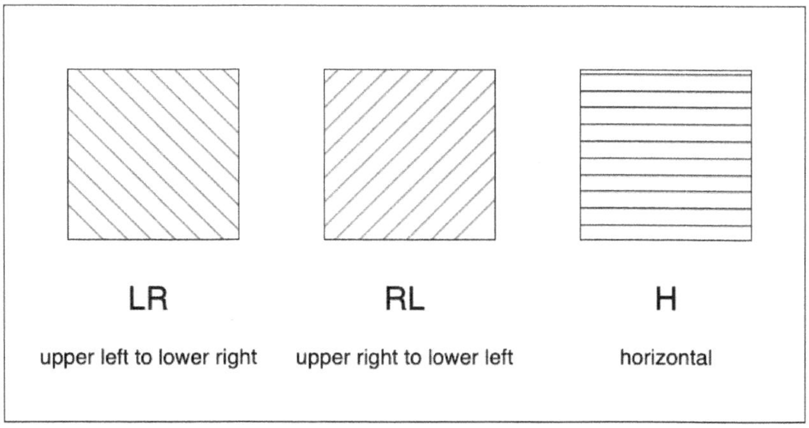

Figure 11.5:

To represent the intersection of two sets - $A \cap B$ - we use the Venn diagram below.

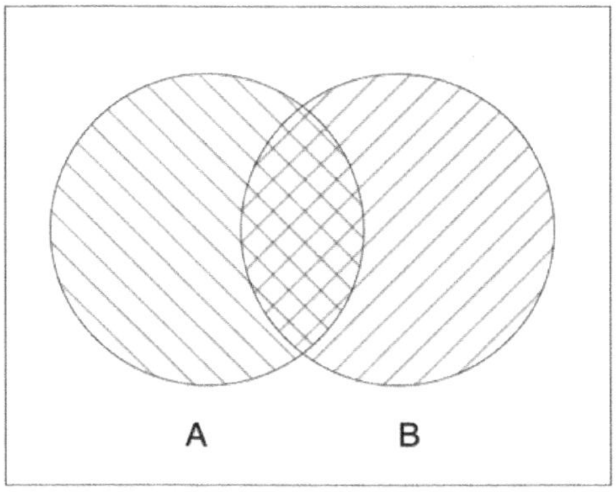

Figure 11.6:

The entire shadowed area represents the union. The LR lines represent all the points of A. The RL lines represent all the points of B. The cross-hatched area - with both LR and RL lines - represents points common to both A and B, and hence points in $A \cap B$.

11.2.2 Venn diagram of $\mathbf{A \cup B}$

To represent the union of two sets - $A \cup B$ - we may also use figure 11.6 above. The union is the sum of the areas with only LR lines plus the area with only RL lines but neglecting repeats. This means points in A and points in B, neglecting repeats.

11.3 distributive laws - intersections and unions

11.3.1 $\mathbf{A} \cup (\mathbf{B} \cap \mathbf{C}) = (\mathbf{A} \cup \mathbf{B}) \cap (\mathbf{A} \cup \mathbf{C})$

This is a **set distributive law**. We now use Venn diagrams to prove the equality in the header 11.3.1.

The Venn diagram for $A \cup (B \cap C)$ (a union) is represented by the combination of **all** the shadowed areas of the figure below, which treats the general case where A, B, C mutually intersect. All shadowed areas are included: the LR-only region, the RL-only region and the cross-hatched region.

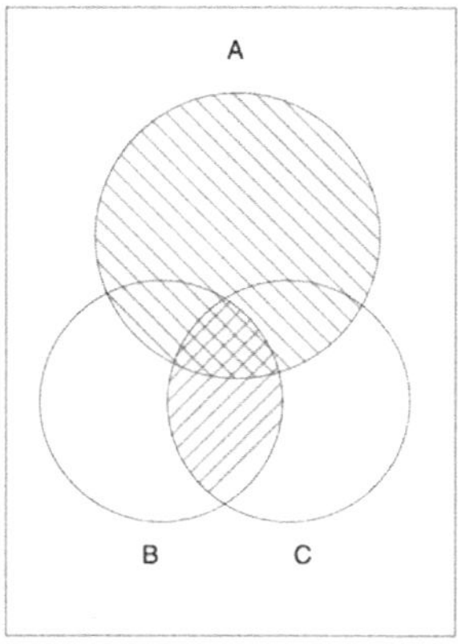

Figure 11.7:

The Venn diagram for $(A \cup B) \cap (A \cup C)$ is given below, by the area containing **only** cross-hatched lines (an intersection). $A \cup B$ is LR lines and $A \cup C$ is RL lines.

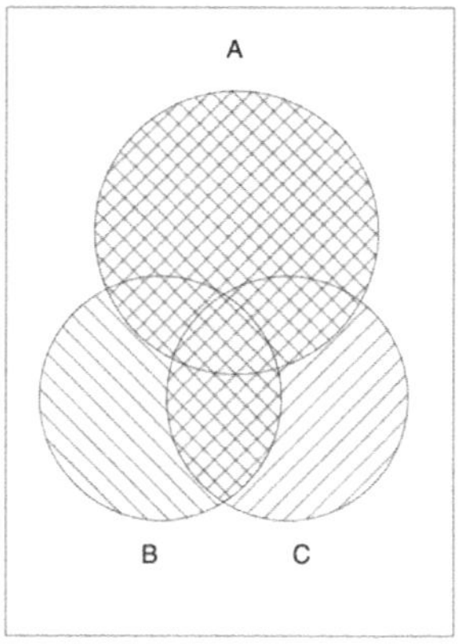

Figure 11.8:

The totally of shadowed areas of figure 11.7 (a union) is the same as the cross-hatched area of 11.8 (an intersection). This is a visual demonstration of the expression in the header 11.3.1.

One can take a combined visual-algebraic approach to this problem as represented in the figure below.

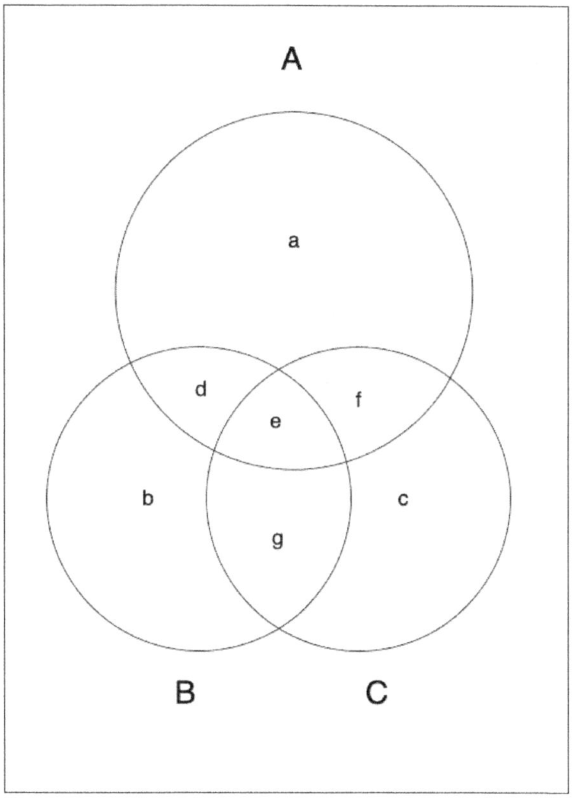

Figure 11.9:

In this case, 'a' is in A but not in B, C, 'd' is in A and B but not C, 'e' is in A, B and C, 'g' is in B and C but not in A, and so forth.

To show the distributive law, header 11.3.1, make a table.

p	$p \in A \cup B$	$p \in A \cup C$	$p \in B \cap C$	$p \in (A \cup B) \cap (A \cup C)$	$p \in A \cup (B \cap C)$
a	Y	Y	N	Y	Y
b	Y	N	N	N	N
c	N	Y	N	N	N
d	Y	Y	N	Y	Y
e	Y	Y	Y	Y	Y
f	Y	Y	N	Y	Y
g	Y	Y	Y	Y	Y

Table 1:

This table should be self-explanatory. The distributive law is valid if the two rightmost columns are the same, which is indeed the case.

11.3.2 $\mathbf{A} \cap (\mathbf{B} \cup \mathbf{C}) = (\mathbf{A} \cap \mathbf{B}) \cup (\mathbf{A} \cap \mathbf{C})$

This is the second **set distributive law** for unions and intersections. In this particular case, the law is similar to an arithmetic distributive law $a * (b + c) = a * b + a * c$, don't read too much into this analogy.

It should be clear that the following Venn diagrams provide a visual demonstration of this expression, header 11.3.2.

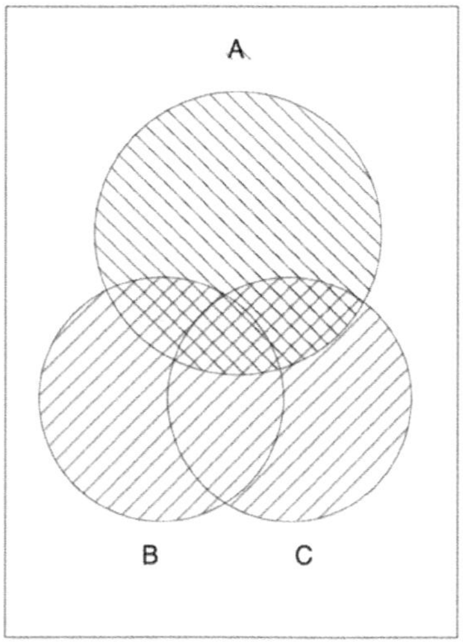

Figure 11.10:

In this figure, the cross-hatchecd lines visually represent the expression $A \cap (B \cup C)$. In the figure below, all the shading together represents the expression $(A \cap B) \cup (A \cap C)$.

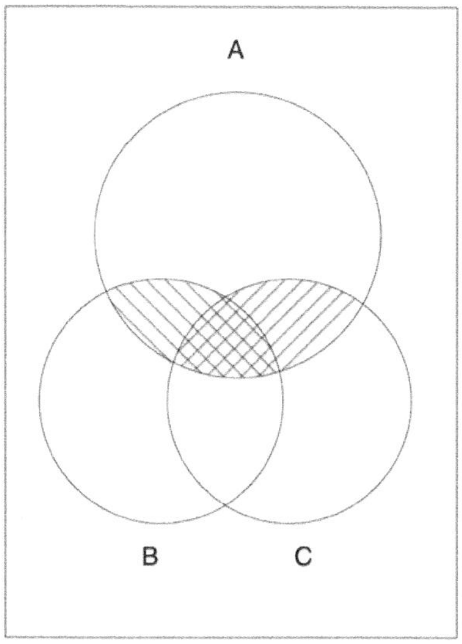

Figure 11.11:

11.4 complements

A **complement** is a third basic operation on sets.

Definition 11.1

Two sets A, B. The ***complement*** *of B relative to A, denoted A/B or A − B is a set*

$$C = A/B \text{ or } C = A - B \tag{11.3}$$

containing all elements of A which are not in B.

If $B \subseteq A$ and if A is implicit, then A/B is also denoted B^c.

The RL lines in the following figure represents two cases of A/B. These lines represent all elements of A which are not in B.

Note that

$$\begin{aligned} A/A &= \emptyset \\ A/\emptyset &= A \\ A/B &= A/(B \cap A) \\ \emptyset/\emptyset &= \emptyset \end{aligned} \tag{11.4}$$

The last is an edge condition. It may or may not have an intuitive interpretation.

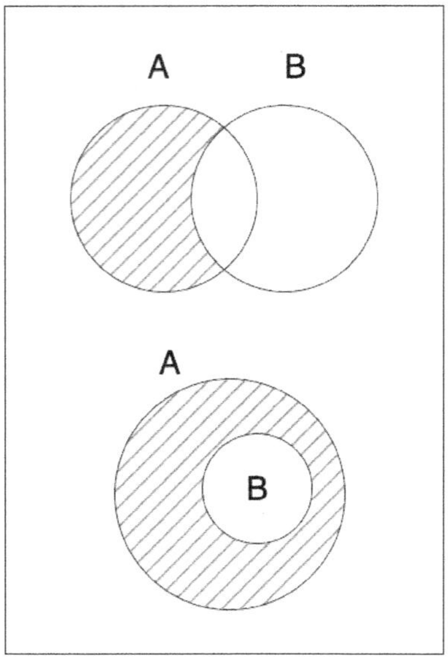

Figure 11.12:

Here are some more examples. Try these before looking at the Venn diagrams below.

$$\begin{aligned} A \cap (A/B) &= A/B \\ A \cup (A/B) &= A \\ B \cup (A/B) &= A \cup B \\ B \cap (A/B) &= \emptyset \end{aligned} \tag{11.5}$$

The cross-hatched lines in the figure below represent $A \cap (A/B)$

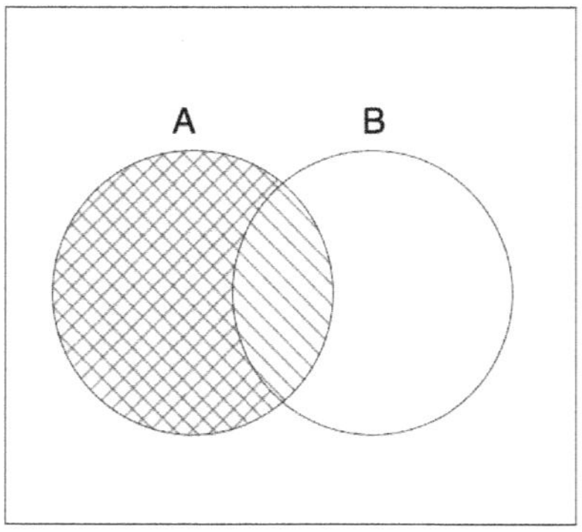

Figure 11.13:

In the figure below, all the shaded areas together represent $B \cup (A/B)$.

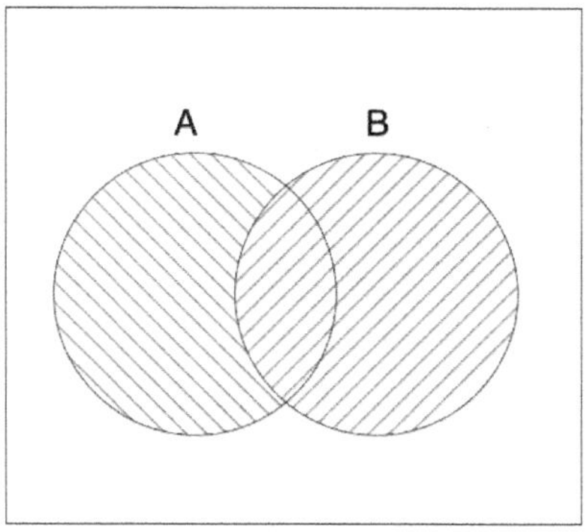

Figure 11.14:

11.5 DeMorgan's laws

DeMorgan's laws are set-valued-expressions involving complements along with unions and intersections. For DeMorgan's laws, we have three sets - A, B, C - such that

$$B, C \subseteq A \tag{11.6}$$

A can be implicit. DeMorgan's laws are a kind of **distributive law**.

11.5.1 $\mathbf{A/(B \cup C) = (A/B) \cap (A/C)}$

If A is implicit, this is

$$(B \cup C)^c = B^c \cap C^c \tag{11.7}$$

Complements are all relative to A. Equation 11.6 holds. The following are the Venn diagrams for this expression. In the top figure the LR lines represent $A/(B \cup C)$. The cross-hatched lines of the bottom figure represent $(A/B) \cap (A/C)$ (the LR lines are A/B, the RL lines are A/C, the cross-hatched area is the intersection). The LR lines of the top figure are the same as the cross-hatched lines of the bottom figure.

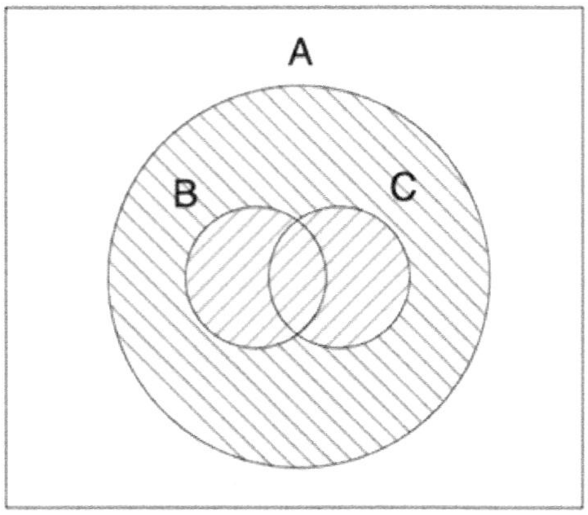

Figure 11.15:

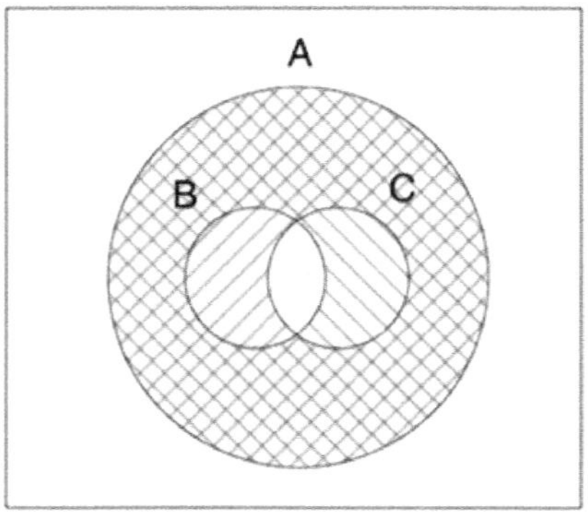

Figure 11.16:

11.5.2 $\mathbf{A}/(\mathbf{B} \cap \mathbf{C}) = (\mathbf{A}/\mathbf{B}) \cup (\mathbf{A}/\mathbf{C})$

If A is implicit, this is

$$(B \cap C)^c = B^c \cup C^c \tag{11.8}$$

The following figures represent this equality. In the top figure, the LR lines represent the left side. In the bottom figure, the entire shaded area represents the right side.

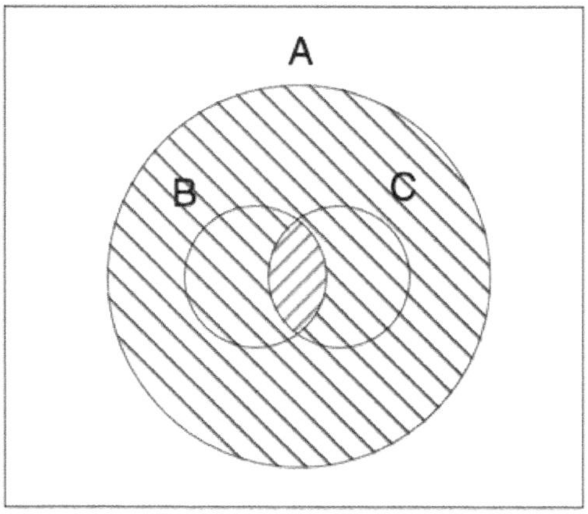

Figure 11.17:

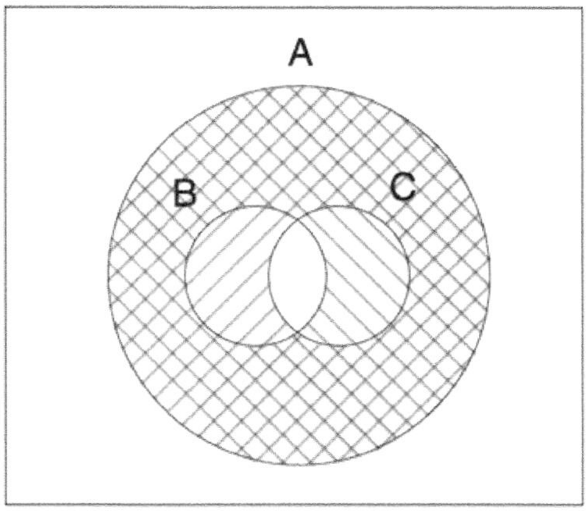

Figure 11.18:

11.6 maps of intersections and unions

This section considers maps whose arguments are unions or intersections. The results of this section are widely used in many areas of mathematics. The following relations, which have been previously demonstrated, will be useful in proving the results. Capital letters represent sets, lower case letters are elements of sets and f() is a function.

$$A \subseteq S \textit{ and } B \subseteq S \Longrightarrow A \cup B \subseteq S \tag{11.9}$$

$$x \in S \Longrightarrow f(x) \in f(S) \tag{11.10}$$

$$A \subseteq S \Longrightarrow f(A) \subseteq f(S) \tag{11.11}$$

$$A \cup A = A, \;\; A \cap A = A \tag{11.12}$$

$$A \subseteq B \textit{ and } B \subseteq A \Longleftrightarrow A = B \tag{11.13}$$

Here is the goal: Given two sets G,H and a map $f : G \to H$, and sets $A, B \in G$, $C, D \in H$, where $C = f(A)$, $D = f(B)$. We are interested in expressions for $f(A \cup B)$, $f(A \cap B)$, $f^{-1}(C \cup D)$, $f^{-1}(C \cap D)$ in terms of functions of A, B, C, D individually.

The general situation is visualized the figure below.

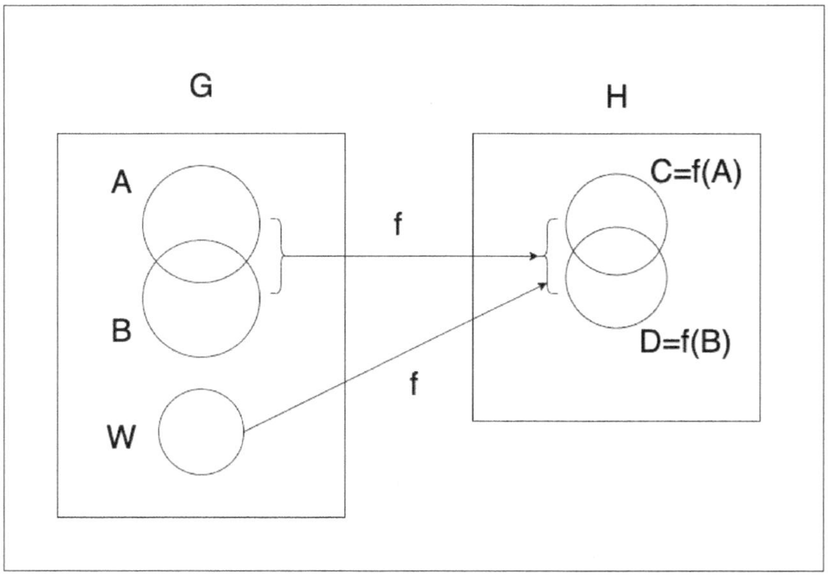

Figure 11.19:

The function f maps set G to set H. In particular, it maps set A to $C = f(A)$ and set B to $D = f(B)$. If f is not injective, then there can be a set $W \subset G$, $W \cap A = W \cap B = \emptyset$ such that $f(W) \subset C \cup D$. In other words, the sets C and D may contain images of elements of G which are not in A or B.

It is assumed that **both A and B are not empty**, otherwise f(A) and/or f(B) would not be defined. It follows that C and D are not empty. Results are the same whether W is empty or not.

11.6.1 $\mathbf{f(A \cup B) = f(A) \cup f(B)}$

The following figure will be useful.

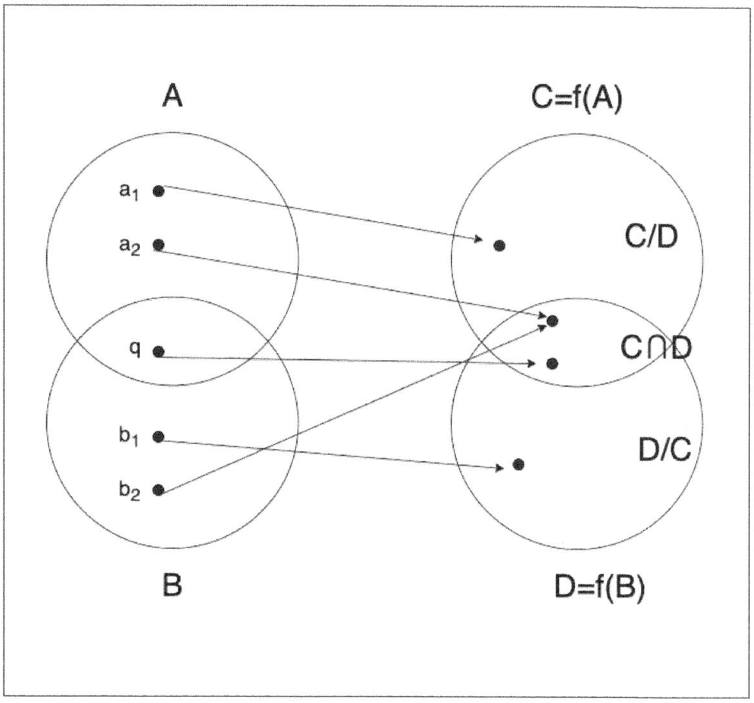

Figure 11.20:

In this figure, elements $a_2 \in A$, $b_2 \in B$ are both mapped to the same element, which must be in $C \cap D$. Similarly, any $q \in A \cap B$ is mapped to $C \cap D$. Therefore $f(A \cup B) = C \cup D$, by definition. But $C \cup D = f(A) \cup f(B)$, also by definition, which proves the expression in the header. As an exercise, show that non-empty W does not affect this result.

11.6.2 $\mathbf{f(A \cap B) \subseteq f(A) \cap f(B)}$

Note subset, not equality. We must assume that $A \cap B$ is not empty, otherwise $f(A \cap B)$ is not defined.

From figure 11.20 it is clear that, for any $q \in A \cap B$, then $f(q) \in C \cap D$. That is, $f(q) \in C$ **and** $f(q) \in D$. Therefore $f(A \cap B) \subseteq C \cap D = f(A) \cap f(B)$.

On the other hand, as seen in figure 11.20, there can be elements in $f(A) \cap f(B)$ (such as $f(a_2) = f(b_2)$) which are not in $f(A \cap B)$ ($a_2 \neq b_2$). Therefore we **cannot** conclude that $f(A) \cap f(B) \subseteq f(A \cap B)$.

11.6.3 $\mathbf{f^{-1}(C \cup D) = f^{-1}(C) \cup f^{-1}(D)}$

Assume both $f^{-1}(C)$, $f^{-1}(D)$ are non-empty. Using figure 11.19, $f^{-1}(C \cup D) = A \cup B \cup W$. Now by definition of pre-image, $f^{-1}(C) = A \cup W_1$ and $f^{-1}(D) = B \cup W_2$, where $W_1 \cup W_2 = W$. Therefore $f^{-1}(C \cup D) = A \cup B \cup W = (A \cup W_1) \cup (B \cup W_2) = f^{-1}(C) \cup f^{-1}(D)$.

11.6.4 $\mathbf{f^{-1}(C \cap D) = f^{-1}(C) \cap f^{-1}(D)}$

Considering figures 11.19 and 11.20, let $A_1 \subseteq A$ be the set of elements of A mapped to $C/(C \cap D)(a_1 \in A_1,\ a_2, q, b_1, b_2 \notin A_1)$ and similarly for A_2, B_1, B_2 and $Q \subseteq A \cap B(q \in Q)$. The sets A_1, A_2, B_1, B_2, Q are mutually disjoint.

Let the set $W_0 \subseteq W$ be such that $f(W_0) \subset (C \cap D)$ and $W_1, W_2 \subseteq W$ be such that $f(W_1) \subset C,\ f(W_2) \subset D$, but $f(W_1) \not\subset C \cap D$ and $f(W_2) \not\subset C \cap D$. The sets W_1, W_2, W_0 are mutually disjoint.

Any A, B or Q set is disjoint to any W set.

Then $f^{-1}(C \cap D) = Q \cup A_2 \cup B_2 \cup W_0$. We also have= $f^{-1}(C) = Q \cup A_1 \cup A_2 \cup B_2 \cup W_1 \cup W_0$ and $f^{-1}(D) = Q \cup B_1 \cup B_2 \cup A_2 \cup Q \cup W_2 \cup W_0$. Therefore, using the distributive laws, $f^{-1}(C) \cap f^{-1}(D) = Q \cup W_0 \cup A_2 \cup B_2 = f^{-1}(C \cap D)$. (As an exercise, work through the details.)

12 equivalence relations and categorization

12.1 introduction

It might be useful to begin with a non-numerical example. Consider the set

$$S = \{apple, chair, peas, table, lamp, sausage\} \tag{12.1}$$

We choose to let the elements of the set S fall into two categories: furniture and food. We define a binary relation on elements of the set, with the symbol $\sim$ as follows:

Definition 12.1

In set theory, the symbol $\sim$ is an ***equivalence relation****, and is read:* ***is equivalent to****. For set S, let $a, b \in S$.*

The notation $a \sim b$ is read: ***a is equivalent to b****. This has the following intuitive interpretations:*

- *a is in the same category as b*
- *a has similar features as b*
- *a is classified the same as b*

So if we write *apple* $\sim$ *peas*, this means 'apple is in the same category as peas' - namely, food - or 'apple is classified the same as peas' - namely food. Similarly for *apple* $\sim$ *sausage*.

In set theory we would say 'apple *is equivalent to* peas' and 'apple *is equivalent to* sausage'. The equivalence is in the context of the particular categories *food, furniture.*

Similarly we have (for set S) *chair* $\sim$ *table*, *chair* $\sim$ *lamp*, *lamp* $\sim$ *chair*. 'chair is classified the same as table', 'lamp is in the same category as chair' - namely, furniture - and so forth.

But we would **NOT** say *apple* $\sim$ *chair*, because, *for the categories we have chosen* - food and furniture - these objects are not categorized the same. **They are not equivalent**.

The classification names are NOT a formal part of the equivalence relation.They may be implicit. See next section.

As another example, consider the set $\mathbb{Z}_+$. If we define two categories *even, odd,* then we would have

$$\begin{aligned} &2 \sim 4 \sim 6 \sim 8...., 2 \sim 6538, 10 \sim 4.... \\ &1 \sim 3 \sim 5 \sim 7...., 183 \sim 45, ... \end{aligned} \tag{12.2}$$

but **not** $1 \sim 20, 7 \sim 4022$ (notated $1 \not\sim 20$, $7 \not\sim 4022$) and so forth. So *for the chosen categories*, 2 is equivalent to 4, 4 is equivalent to 9786, 3 is equivalent to 67 and so forth. Any even (odd) number is equivalent to any other even (odd) number.

12.2 equivalence relation, equivalence classes, partition

We now formalize and extend the above discussion.

Definition 12.2

Given set S, we define an ***equivalence relation*** $\sim$ *by specifying which elements are equivalent to which other elements. For any* $a, b, c \in S$, *the notation* $a \sim b$ *means* ***a is equivalent to b****.*

The equivalence relation $\sim$ *has the following properties.*

1. $a \sim a$, *(symmetric)*
2. $a \sim b$ *implies* $b \sim a$, *(reflexive)*
3. *if* $a \sim b$ *and* $b \sim c$ *then* $a \sim c$, *(transitive)*

Each element of S must be included in an equivalence relation. It is possible that one or more elements of S are equivalent only to themselves (that is, only item 1 applies).

Our intuitive understanding of the above list is:

1. Any element a is in the same category as itself.
2. If b is in the same category as a, then a is in the same category as b.
3. If a, b are in the same category and b, c are in the same category, then a, c are in the same category.

Equivalently we could use the words 'classification' or 'classify' rather than 'categorization' or 'category':

1. Any element a is has the same classification as as itself.
2. If b has the same classification as a, then a has the same classification as b.

3. If a, b are classified the same and b, c are classified the same, then a, c have the same classification..

The symbol $\sim$ itself is called the 'equivalence relation'.

An equivalence relation provides information about the elements of a set - namely, how some elements are the same or different in a particular way. It puts a structure on the set and, as will be seen, this structure has interesting properties.

For a finite set the maximum number of categories equals the number of elements in the set. For an infinite set there can be an infinite number of categories. For example, the real-number intervals [0,1), [1,2), [2,3),.... of $\mathbb{R}$ may be different categories.

It is possible that each category has only one member. This is not necessarily trivial. It is possible that some but not all categories have only one member. It is normal (but not required) that all categories have more than one member. It is also possible to define a trivial equivalence relation where all the elements of the set are equivalent to each other.

An equivalence relation $a \sim b$ between two elements of a set is equivalent to a symmetric ordered pair $(a, b) = (b, a)$.

> From this point **equivalence relations will be treated abstractly**. That is, we will not justify an equivalence relation with some intuitive classification scheme. This will make it easier to discuss the mathematical structure of equivalence relations.

What do we do with the elements of a set which are equivalent to each other? We put them in their own set. Here is an example.

$$\begin{aligned} &S = \{a, b, c, d, e, f\} \\ &a \sim b \sim c \\ &d \sim e \\ &f \sim f \end{aligned} \tag{12.3}$$

There are 3 categories, so we create 3 sets, called **equivalence classes** (recall that *class* is a synonym for *set*), one set (or equivalence class) for each group of elements which are all equivalent to each other. Since sets are typically indicated by capital letters, we might indicate equivalence classes as

$$\begin{aligned} X_a &= \{a, b, c\} \\ X_d &= \{d, e\} \\ X_f &= \{f\} \end{aligned} \tag{12.4}$$

A **representative** of a non-empty set is some element from the set. Any element can be a representative. The idea is that, in some circumstances, the representative of a set stands in for the entire set.

An equivalence class is typically labeled by a **representative** of the set, so equivalence class X_a is labeled by 'a', which is a representative of the set $\{a, b, c\}$. Any representative could be used so

$$\begin{aligned} X_a = X_b = X_c = \{a, b, c\} \\ X_d = X_e = \{d, e\} \end{aligned} \tag{12.5}$$

The label of equivalence class $X_f = \{f\}$ is unique as f is the only representative.

There is a very common notation for an equivalence class. The equivalence class containing element 'a' is denoted **[a]**. Therefore, as an alternative to the above we would write

$$\begin{aligned} [a] = [b] = [c] = \{a, b, c\} \\ [d] = [e] = \{d, e\} \\ [f] = \{f\} \end{aligned} \tag{12.6}$$

The equivalence classes of even, odd numbers of $\mathbb{Z}_+$ could be labeled, for example, $N_{even} = \{2, 4, 6, 8, ...\}$, $N_{odd} = \{1, 3, 5, 7, ...\}$ as well as, weirdly, [28], [37],, etc. A notation like [even], [odd], is not used, because the tokens 'even' and 'odd' are not representatives of a set.

Definition 12.3

Given non-empty set S with an equivalence relation $\sim$, the ***set*** *of all elements of S which are equivalent to each other is called an* ***equivalence class****.*

Two elements $a, b \in S$ are in the same equivalence class if and only if $a \sim b$.

Intuitively, all the elements in an equivalence class have the same classification or category.

Every element of S must be in some equivalence class. *No equivalence class is empty.*

The last part of the above definition is important. If equivalence classes are to be defined on a set, then every element of the set must be in some equivalence class, and every equivalence class must contain at least one element of the set.

Here are a few examples.

In $\mathbb{R}$, equivalence classes may be defined by the following sets: $(-\infty, 0), [0, 1), [1, 2), [2, 3), ...$

A **dictionary** spatially embodies an equivalence relation. All words starting with 'a' are in successive pages, all words that start with 'b' are in the next group of pages and so forth.

Two **football teams** just before the ball is snapped embody an equivalence relation for 'side of the line of scrimmage'.

Consider the **set of all closed intervals of the real line**. An equivalence relation might be: Two intervals [a,b] and [c,d] are equivalent - $[a, b] \sim [c, d]$ - when b - a = d - c. That is, when they have the same length. Therefore we might notate the equivalence classes as L_d, where d is the length of the interval. L_d is the set of all closed intervals of length d. Note that this is not equivalent to the notation [d], which is meaningless in this context because d is not a representative of an interval. On the other hand, if q = p + d, then

$$[\,[p, q]\,] = L_d = \{[a, b], [a', b', [a'', b''], ...\}, \quad b - a = b' - a' = b'' - a''.... = d \tag{12.7}$$

Consider the **set of all points (x,y) of the plane**. Two points may be defined as equivalent if they lie on the same circle centered at the origin. Therefore any equivalence class can be represented by the radius R of the circle - C_R - this is the set of all points (x,y) on the circle of radius R and centered at the origin. If we try to define equivalence based on points lying on a square, the square would have to be centered somewhere as otherwise two squares might intersect and a point might be in two equivalence classes, meaning the definition of equivalence is not valid.

The set of equivalence classes of some set S defines a **partition** of the set S. Partitions were discussed in section 7.6. The set of equivalence classes, equations 12.5 or 12.6

$$\{X_a, X_d, X_f\} = \{[b], [e], [f]\} \tag{12.8}$$

form a partition of set S of equation 12.3.

Theorem 12.4

An equivalence relation defines a partition. *The partition is the collection of equivalence classes.*

The set of equivalence classes is the same as a partition. For an equivalence relation on set S, an element of S must be in one and only one equivalence class and every element of S must be in some equivalence class. In a partition of S, every element of S must be in one and only one set of the partition, and every element of S must be in some set of the partition. Therefore a set of a partition is identical to an equivalence class.

Theorem 12.5

A partition defines an equivalence relation. *Given a partition, there exists an equivalence relation whose equivalence classes are the sets of the partition.*

For example, given set $S = \{1, 2, 3, 4, 5, 6, 7, 8, 9, 10\}$ and a partition $P = \{\ \{1, 3, 4, 6\}, \{2, 5\}, \{7, 8, 9, 10\}\ \}$, the corresponding equivalence relation is:

$$\begin{aligned} &1 \sim 3 \sim 4 \sim 6 \\ &2 \sim 5 \\ &7 \sim 8 \sim 9 \sim 10 \end{aligned} \tag{12.9}$$

The sets of the partition P are equivalence classes

$$[6] = X_6 = \{1, 3, 4, 6\}, \quad [5] = X_5 = \{2, 5\}, \quad [7] = X_7 = \{7, 8, 9, 10\} \tag{12.10}$$

12.3 equivalence relations and surjective maps

We have seen that a surjective map corresponds to a partition and vice versa (section 7.6). Since a partition corresponds to an equivalence relation, it follows that **a surjective map corresponds to an equivalence relation**.

The procedure for creating an equivalence relation based on a surjective map is illustrated by example. Consider two sets

$$S = \{1, 2, 3, ..., 10\}, \quad T = \{e, o\} \tag{12.11}$$

Define a surjective map F

$$\begin{aligned} &F : S \to T \\ &F(s) = \begin{cases} e & s\ even \\ o & s\ odd \end{cases} \\ &F(1) = F(3) = ... = o, \quad F(2) = F(4) = ... = e \end{aligned} \tag{12.12}$$

F is surjective because $T = im(F)$. It is not injective because there exists (at least) two elements s, s' of S such that $F(s) = F(s')$. In this case, F^{-1} is a *pre-image* and not an inverse.

We may now create equivalence classes and an equivalence relation from the pre-image of F.

$$[1] = F^{-1}(o) = \{1, 3, 5, 7, 9\}, \quad [2] = F^{-1}(e) = \{2, 4, 6, 8, 10\} \tag{12.13}$$

As discussed in section 7.6, the pre-image of a surjective map is a partition. **The sets of F^{-1} partition S**. This is easily seen in equation 12.13. The partition P is the collection of sets

$$P = \{[1], [2]\} \tag{12.14}$$

The elements of P are the equivalence classes. The equivalence relations are deduced from the partition.

$$1 \sim 3 \sim 5 \sim 7 \sim 9, \qquad 2 \sim 4 \sim 6 \sim 8 \sim 10 \tag{12.15}$$

This example generalizes:

Theorem 12.6

Given nonempty sets S, T and surjective map $F : S \rightarrow T$. For each element $t \in T$, the set

$$\Omega_t = F^{-1}(t) \tag{12.16}$$

is an equivalence class of S. (Note that t is not a representative of the equivalence class but is used for notation.)

The set of equivalence classes

$$P = \{F^{-1}(t)|\forall t \in T\} = \{\Omega_t|\forall\, t \in T\} \tag{12.17}$$

partitions S. Every $s \in S$ is in some Ω_t. Also

$$\Omega_t \cap \Omega_{t'} = \emptyset \text{ when } t \neq t' \tag{12.18}$$

For any $t \in T$, all the elements of Ω_t are equivalent.

$$If\ t' \in \Omega_t,\ then\ t \sim t' \tag{12.19}$$

For proof, refer to section 7.6. Exercise: What are the equivalence classes if F is bijective?

Note that if $p \in \Omega_t,\ q \in \Omega_{t'},\ t \neq t'$ then $p \nsim q$. That is, p is **not equivalent** to q.

If there exists a map $F : S \rightarrow T$ which is not surjective, replace T in the theorem with $T \rightarrow im(F)$. The map $F : S \rightarrow im(F)$ is surjective.

We can turn things around, to **define a surjective map based on an existing equivalence relation**.

Theorem 12.7

Given set S, equivalence relation $\sim$ *and equivalence classes* $E_1, E_2, ..., E_N$*. (For simplicity, we assume a finite number of equivalence classes, but this is not a necessary restriction.) Then we may define a set (essentially an index set, with subscripts of the equivalence classes)*

$$T = \{1, 2, 3, ..., N\} \tag{12.20}$$

and a surjective map

$$\begin{aligned} &F : S \to T \\ &F(e) = i \ \ when \ e \in E_i \end{aligned} \tag{12.21}$$

The map F reproduces the existing equivalence classes

$$F^{-1}(i) = E_i \tag{12.22}$$

If the set of equivalence classes is countably infinite or uncountable, then let α be the index set for the equivalence classes, and instead of equation 12.20 we would have

$$T = \alpha \tag{12.23}$$

Otherwise everything goes through as is. In short

A surjective map defines an equivalence relation.
An equivalence relation defines a surjective map.

Below is an example, which uses the set

$$S = \{a, b, c, d, e, f, g, h, i\} \tag{12.24}$$

surjective map from equivalence relation

Define the equivalence relation

$$\begin{aligned} &a \sim b \sim c \sim d \\ &e \sim f \\ &g \sim h \sim i \end{aligned} \tag{12.25}$$

1. Define equivalence classes E_i and index set T

$$E_1 = \{a, b, c, d\}, \quad E_2 = \{e, f\}, \quad E_3 = \{g, h, i\}, \quad T = \{1, 2, 3\} \tag{12.26}$$

2. define surjective map F

$$\begin{aligned}
&F : S \to T \\
&F(s) = k \quad s \in E_k, \quad k = 1, 2, 3 \\
&F(a) = F(b) = F(c) = F(d) = 1 \\
&F(e) = F(f) = 2 \\
&F(g) = F(h) = F(i) = 3 \\
&F(E_1) = \{1\}, \ , F(E_2) = \{2\}, \quad F(E_3) = \{3\}
\end{aligned} \tag{12.27}$$

3. Show F^{-1} reproduces the equivalence classes
   For example, $F^{-1}(2) = E_2$.

equivalence relation from surjective map

Start with set S, the set $T = \{1, 2, 3\}$ and the surjective map $F : S \to T$

$$\begin{aligned}
&F(a) = F(b) = F(c) = F(d) = 1 \\
&F(e) = F(f) = 2 \\
&F(g) = F(h) = F(i) = 3
\end{aligned} \tag{12.28}$$

1. Define sets E_i, which are equivalence classes.

$$E_1 = F^{-1}(1) = \{a, b, c, d\}, \quad E_2 = F^{-1}(2) = \{e, f\}, \quad E_3 = F^{-1}(3) = \{g, h, i\} \tag{12.29}$$

2. Create equivalence relation $\sim$ from the equivalence classes.
   From $E_1 : a \sim b \sim c \sim d$
   From $E_2 : e \sim f$
   From $E_3 : g \sim h \sim i$

12.4 quotient maps, quotient sets

Definition 12.8
The collection (ie, set) of equivalence classes of set S, under equivalence relation $\sim$, is denoted

$$S/_\sim \tag{12.30}$$

*The collection $S/_\sim$ is called a **quotient set**.*

In the case that some collection P of sets partitions S, define $\sim$ from the sets of P and define the quotient set

$$S/_\sim = P \tag{12.31}$$

For the example, equations 12.11 and 12.12,

$$S/_{\sim} = \{[1], [2]\} \tag{12.32}$$

For the example, equation 12.27,

$$S/_{\sim} = \{E_1, E_2, E_3\} \tag{12.33}$$

The elements of $S/_{\sim}$ are sets. The elements of S are individual objects which are elements of the sets of $S/_{\sim}$.

Define the map π

$$\begin{aligned} \pi &: S \to S/_{\sim} \\ \pi(s) &= [s] \end{aligned} \tag{12.34}$$

where $s \in S$ and $[s]$ is the equivalence class containing s. That is, π maps an element of S to its equivalence class in the quotient set $S/_{\sim}$.

This is a map **from the elements of set S to the sets which are elements of the quotient set** $S/\sim$.

Continuing the the example, equations 12.11 and 12.12, the map π is

$$\pi(1) = \pi(3) = ... = \pi(9) = [1], \quad \pi(2) = \pi(4) = ... = \pi(10) = [2] \tag{12.35}$$

This map is called a **quotient map**. Here is the generalization.

Definition 12.9

Given set S with equivalence relation $\sim$ and quotient set $S/_{\sim}$ ($S/_{\sim}$ is a collection of equivalence classes), The map

$$\begin{aligned} \pi &: S \to S/_{\sim} \\ \pi(s) &= [s] \quad \forall\, s \in S \end{aligned} \tag{12.36}$$

*is called a **quotient map**.*

Note that **a quotient map is surjective** because every s in S is in one and only one equivalence class and every equivalence class contains at least one element of S. So π maps S **onto** $S/_{\sim}$. The quotient map is not in general injective.

12.5 universal property. Commuting diagrams

Combining the surjective map $F : S \to T$ and surjective quotient map $\pi : S \to S/_\sim$ (where both correspond to the same partition of S), we have the figure

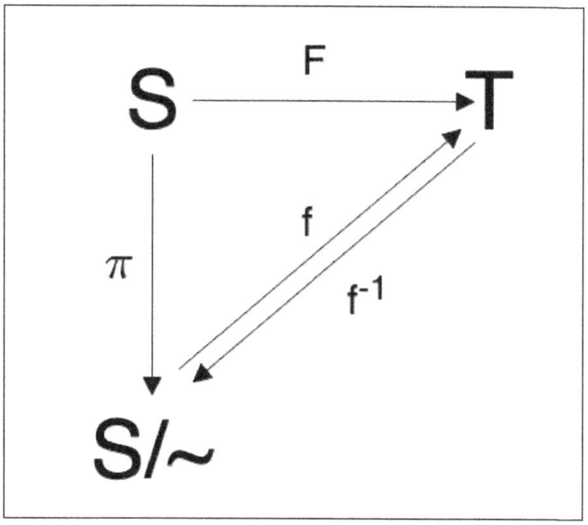

Figure 12.1:

A diagram like this relates sets in many areas of mathematics.

A third map **f** $: \mathbf{S}/\sim\ \to \mathbf{T}$ **is bijective, so the inverse** $\mathbf{f}^{-1} : \mathbf{T} \to \mathbf{S}/\sim$ **is also a bijective map.** The map f is shown in the above figure.

Here is an example.

$$
\begin{aligned}
& S = \{1,2,3,4,5,6\}, \quad T = \{o, e\} \\
& F(1) = F(3) = F(5) = o, \quad F(2) = F(4) = F(6) = e \\
& 1 \sim 3 \sim 5, \quad 2 \sim 4 \sim 6 \\
& [1] = \{1,3,5\}, \quad [2] = \{2,4,6\} \\
& \pi(1) = \pi(3) = \pi(5) = [1], \quad \pi(2) = \pi(4) = \pi(6) = [2] \\
& S/\sim\ = \{[1],[2]\}
\end{aligned}
\tag{12.37}
$$

The situation is illustrated below.

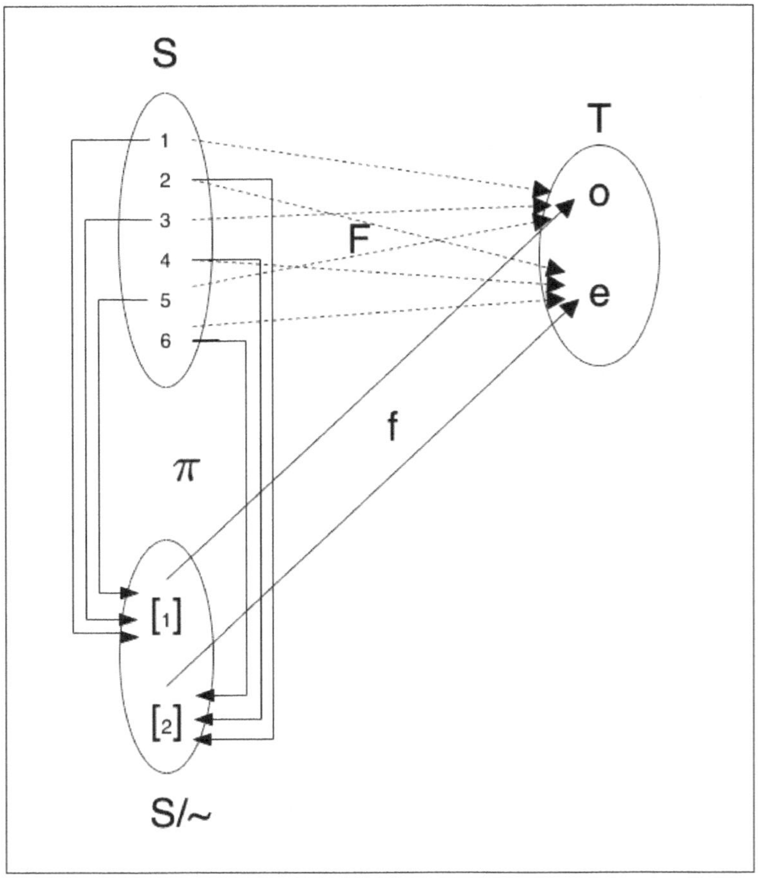

Figure 12.2:

It is clear that the elements of T are in 1-1 correspondence with the elements of $S/\sim$. Although the elements of $S/\sim$ are sets and the elements of T are abstract (or variables), there is a 1-1 correspondence in the following sense: **The functions F and π act similarly: Each maps all the elements of a given equivalence class to a single entity - either to a set in $S/\sim$ or an element of T**. So the elements of T are in 1-1 correspondence with the elements of $S/\sim$, which means f, figure 12.2, is bijective.

In this figure we therefore have

$$\begin{aligned} f([1]) &= F(1) = F(3) = F(5) = o \\ f([2]) &= F(2) = F(4) = F(6) = e \end{aligned} \tag{12.38}$$

In fact, for any S, T and surjective F, such that an equivalence relation is defined (so π is defined), the function f is **defined** such that

$$f([s]) = F(s), \quad f^{-1}(F(s)) = [s] \tag{12.39}$$

which is valid for all $s \in S$.

It follows from equation 12.39 that

$$(f \cdot \pi)(s) = f(\pi(s)) = f([s]) = F(s) \tag{12.40}$$

And since this is valid for all $s \in S$, we may write

$$f \cdot \pi = F \tag{12.41}$$

Considering figure 12.1, the above equation states that the *direct path* from S to T (via F) is equivalent to the *indirect path* from S through $S/\sim$ to T (via $f \cdot \pi$). We say that the **diagram commutes**.

Since f is bijection, we may compose equation 12.41 with f^{-1} from the left (see the end of section 8.5.1) which gives $f^{-1} \cdot f \cdot \pi = f^{-1} \cdot F$, and from equation 8.13 and the associativity of composition

$$\pi = f^{-1} \cdot F \tag{12.42}$$

This is equivalent to equation 12.41 and is an alternative expression of the commutativity of the diagram of figure 12.1. It says that the direct path from S to $S/\sim$ (via π) is equivalent to the indirect path from S through T to $S/\sim$ (via $f^{-1} \cdot F$). This is the other aspect of the commutativity of the diagram.

Any three sets whose structures are isomorphic with S, T and $S/\sim$, with functions equivalent to F, f and π, are said to have the **universal property**. Such structures are found in many areas of mathematics.

13 cartesian product

13.1 cartesian product of two sets

13.1.1 definition-notation

We start with an example. Let

$$A = \{a, b, c\}, \quad B = \{1, 2, 3, 4\} \tag{13.1}$$

The **cartesian product** of A and B, denoted $A \times B$ is a set of ordered pairs (section 7.1)

$$\begin{aligned} A \times B =& \{(a,1),(a,2),(a,3),(a,4),\{b,1),(b,2),\\ &(b,3),(b,4),(c,1),(c,2),(c,3),(c,4)\} \end{aligned} \tag{13.2}$$

The general definition is:

Definition 13.1

Let A,B be two nonempty sets: finite, countably infinite or uncountable. The cartesian product is

$$A \times B = \{(a,b)|a \in A, b \in B\} \tag{13.3}$$

Note that, if $A \neq B$, $A \times B \neq B \times A$.

A notation that will be useful is

$$\{0,1\}^2 = \{0,1\} \times \{0,1\} = \{(0,0),(0,1),(1,0),(1,1)\} \tag{13.4}$$

This set is isomorphic to the set of numbers 0,1,2,3 in binary representation: 00,10,01,11.

For another example, let $A = \{0, 1, 2, ..., 9\}$, $A^* = A/\{0\} = \{1, 2, ..., 9\}$. Then

$$A^* \times A = \{(1,0),(1,1),...,(1,9),\ (2,0),(2,1),...,(2,9),...,(9,9)\} \tag{13.5}$$

which is isomorphic to all the integers 10,11,...,99.

Either set may have only one element.

$$\{1\} \times A = \{(1,a),(1,b),(1,c)\}, \quad A \times \{1\} = \{(a,1),(b,1),(c,1)\} \tag{13.6}$$

The product

$$\{1\} \times \mathbb{Z}_+ = \{(1,1),(1,2),(1,3),....\} \tag{13.7}$$

is isomorphic to $\mathbb{Z}_+$. For any $n \in \mathbb{Z}_+$, the ordered pair =(1,n) is in the set $\{1\} \times \mathbb{Z}_+$.

The cartesian product may be written as a union:

$$A \times B = \bigcup_{x \in A} (\{x\} \times B) = (\{a\} \times B) \cup (\{b\} \times B) \cup (\{c\} \times B) \tag{13.8}$$

Similarly

$$\begin{aligned} A \times B &= \bigcup_{y \in B} A \times \{y\} \\ A \times B &= \bigcup_{x \in A} \bigcup_{y \in B} \{(x, y)\} \end{aligned} \tag{13.9}$$

Exercise: Illustrate the unions using the example equation 13.1

It is not uncommon to take the cartesian product of a set with itself. For the set A of equation 13.1

$$A^2 = A \times A = \{(a,a),(a,b),(a,c),(b,a),(b,b),(b,c),(c,a),(c,b),(c,c)\} \tag{13.10}$$

For any non-empty set A, the cartesian products $A \times \emptyset$, $\emptyset \times A$, $\emptyset \times \emptyset$ are undefined.

Exercise: write the cartesian product of two infinite sets in terms of index sets.

13.1.2 disjoint union

A disjoint union of two sets is a union in which **all** elements of both sets are in the final union. This means that what would be repeated elements in an ordinary union are distinguished in a disjoint union.

Definition 13.2

Given two non-empty sets $A = \{a_1, a_2, ...\}, B = \{b_1, b_2, ...\}$, *the* ***disjoint union*** *of the sets is*

$$A \sqcup B = (\{0\} \times A) \cup (\{1\} \times B) = \{(0, a_1), (0, a_2), ..., (1, b_1), (1, b_2), ...\} \tag{13.11}$$

The generalization to any finite number of finite or infinite sets is straightforward.

13.1.3 cardinality

Theorem 13.3

For finite nonempty sets A, B the cardinality of $A \times B$, denoted $|A \times B|$, is

$$|A \times B| = |A|\,|B| \tag{13.12}$$

For example, for equations 13.1, $|A \times B| = 3 * 4 = 12$.

Theorem 13.4

For set A nonempty and finite or countably infinite, the cartesian product $A \times \mathbb{Z}_+$ is countably infinite.

The corresponding union is either a finite or countably infinite union of countably infinite sets (equations 13.8 and 13.9), which is countably infinite (sections 8.4.4 and 8.4.5).

Theorem 13.5

Let A be a finite nonempty set and B be an uncountable set ($|B| = \aleph_1$) Then $A \times B$ is uncountable ($|A \times B| = \aleph_1$).

Let Γ be the uncountable index set for elements of B. That is, there is a bijection $h : \Gamma \to B$ so that $B = \{h(\gamma)|\gamma \in \Gamma\}$. We cannot write something like $B = \{b_1, b_2, ...\}$ as this would imply B is countably infinite rather than uncountable.

If D is any other set such that there is a bijection $J : \Gamma \to D$, then D is uncountable and $D \cong B$. There is a bijection $G : B \to D$, $G = J \cdot h^{-1}$ from B to D (via the intermediary set Γ).

Let $A = \{1\}$ and B as above. The set

$$S = A \times B = \{(1, b)|b \in B\} \tag{13.13}$$

is uncountable because there is the bijection K

$$K : B \to S, \quad K(b) = (1, b)\ \forall b \in B \tag{13.14}$$

Now consider $A = \{1, 2\}$ so that

$$A \times B = \{(1, b)|b \in B\} \cup \{(2, b)|b \in B\} \tag{13.15}$$

$A \times B$ is uncountable because it is the union of two disjoint uncountable sets. That is,

$$|A \times B| = \aleph_1 \tag{13.16}$$

This generalizes to any finite set A. Now suppose A is countably infinite.

Theorem 13.6

Let set A be countably infinite and B uncountable, Then $A \times B$ is uncountable - that is, $|A \times B| = \aleph_1$.

We can let $B = [0, 1)$ with no loss of generality (why?). Since A is countably infinite, we may denote $A = \{a_1, a_2, a_3, ...\}$ (bijection with arbitrary A implied). Then

$$\{a_i\} \times B \cong \{a_i\} \times [0, 1) \cong \{a_i\} \times [n, n + 1) \cong \{1\} \times [n, n + 1) \tag{13.17}$$

for **any** $n \in \mathbb{Z}_+$. The last step is the important one. Then

$$\begin{aligned} A \times B \cong A \times [0, 1) &= \cup_{n \in \mathbb{Z}_+} \{a_n\} \times [0, 1) \\ &\cong \cup_{n \in \mathbb{Z}_+} \{a_n\} \times [n, n + 1) \\ &\cong \{1\} \times \cup_{n \in \mathbb{Z}_+} [n, n + 1) \\ &\cong \{1\} \times \mathbb{R}_+ \cong \mathbb{R}_+ \end{aligned} \tag{13.18}$$

and therefore $|A \times B| = \aleph_1$.

13.1.4 $\mathbb{R} \cong \mathbb{R}^2$, the two-dimensional plane

A point of the two-dimensional plane, in cartesian coordinates, is notated (x,y), an ordered pair of real numbers. The entire two-dimensional plane is denoted $\mathbb{R}^2$, which is the cartesian product of $\mathbb{R}$ with itself

$$\mathbb{R}^2 = \mathbb{R} \times \mathbb{R} = \{(x, y)|x, y \in \mathbb{R}\} \tag{13.19}$$

A grid in the 2-D plane is an ordered pair of integers (i,j). It is is denoted $\mathbb{Z}^2$, which is the cartesian product of $\mathbb{Z}$ with itself

$$\mathbb{Z}^2 = \mathbb{Z} \times \mathbb{Z} = \{(i, j)|i, j \in \mathbb{Z}\} \tag{13.20}$$

Just as $\mathbb{Z}^2$ is countably infinite (theorem 13.4), we have

Theorem 13.7

$\mathbb{R}^2$ *is uncountable. That is,* $|\mathbb{R}|^2 = \aleph_1$.

This theorem states that **the cardinality of the cartesian plane equals the cardinality of the real line.** In other words, there is a bijection between $\mathbb{R}^2$ and $\mathbb{R}$.

Georg Cantor spent three years trying to prove that this theorem is false. When he finally did prove the theorem, he was apparently quite surprised. The following is the essence of Cantor's proof.

First prove the following:

$$(0,1)^2 \cong (0,1) \tag{13.21}$$

That is, the open unit square $(0,1)^2$ is isomorphic to the open interval (0,1).

We assume the binary expansion of the real numbers, with only terminating 0's and not terminating 1's. Let X, Y be sets of all real numbers. We will denote the numbers as

$$X : 0.x_1x_2x_3.... \quad Y : 0.y_1y_2y_3.... \tag{13.22}$$

the real numbers of the two real lines, where $x_i, y_i\ in\{0,1\}$. Equality x = y iff $x_i = y_i$ for all $i \in \mathbb{Z}_+$.

For **any** two real numbers x=$0.x_1x_2x_3...,\ , y = 0.y_1y_2y_3...in(0,1)$, define the real number $z = 0.x_1y_1x_2y_2x_3y_3.....$ whose digits come alternately from x and y. There is a 1-1 correspondence between the pair (x,y), a point in the plane, and the number z.

First suppose that (x,y) and (x',y') give the same z, but $x \neq x'$ or $y \neq y'$ or both. Suppose $x \neq x'$. Then, for at least one n, $x_n \neq x'_n$. This means that the digit z_{2n-1} equals both x_n and x'_n, which is a contradiction. The same argument holds for the digits of y.

Now suppose that both z and z' correspond to the same pair (x,y) but $z \neq z'$. But then for some n, $z_n \neq z'_n$ which means some digit of x (or y) must be different in the same x (or y), a contradiction.

The reason this proof works, of course, is because two countably infinite strings of digits x and y can be alternately merged to a countably infinite string of digits z. This is the great flexibility of infinity, which expands or contracts as needed.

This establishes equation 13.21. To establish theorem 13.7 is straightforward. We have

$$(0,1)^2 = \bigcup_{\alpha,\beta\in(0,1)} \{(\alpha,\beta)\} \tag{13.23}$$

Since, as has been shown, there is a bijection between (0,1) and $\mathbb{R}$, then using definition 13.1 and equation 13.18

$$\begin{aligned} \mathbb{R} \cong (0,1) \cong (0,1)^2 = \bigcup_{\alpha,\beta\in(0,1)} \{(\alpha,\beta)\} \\ \cong \bigcup_{\alpha\in\mathbb{R},\beta\in(0,1)} \{(\alpha,\beta)\} \cong \bigcup_{\alpha\in\mathbb{R},\beta\in\mathbb{R}} \{(\alpha,\beta)\} = \mathbb{R}^2 \end{aligned} \tag{13.24}$$

Exercise: Fill in the steps explicitly in going from the first line to the second line in the above equation.

13.2 cartesian product of multiple sets

13.2.1 notation

Consider finite sets

$$A = \{a_1, a_2, a_3\}, \quad B = \{b_1, b_2\}, \quad C = \{c_1, c_2, c_3, c_4\} \tag{13.25}$$

The cartesian product is

$$A \times B \times C = \{(a_1,b_1,c_1),(a_2,b_1,c_1),(a_3,b_1,c_1),(a_1,b_2,c_1),(a_1,b_1,c_2),....,(a_3,b_2,c_4)\} \tag{13.26}$$

is a set of **3-tuples** with 3 **slots**. A tuple with N elements, $(x_1, x_2, ..., x_N)$ is an **N-tuple** with N **slots**. Tuples can also have a countably infinite number of slots or an uncountable number of slots as well.

There is a very useful notation for cartesian products. Let I be an index set which may be finite, uncountably infinite or uncountable, and let $\{A_\alpha\}_{\alpha\in I}$ be a set of sets. Then the cartesian product of all the sets A_α may be notated

$$A_\alpha \times A_{\alpha'} \times A_{\alpha''} \times ... = \prod_{\alpha\in I} A_\alpha \tag{13.27}$$

13.2.2 countably infinite product of finite sets is uncountable

The countably infinite cartesian product of the set $\{0,1\}$ is uncountable. That is

Theorem 13.8

$$\{0,1\} \times \{0,1\} \times \{0,1\} \times .... = \prod_{i=1}^{\infty} \{0,1\} \cong \mathbb{R} \tag{13.28}$$

The cartesian product has the form $\{(1,0,0,1,1,0,1,...)\}, \{(0,0,1,1,1,1,0,1,0,1,...)\}, ...\}$, an infinite set of infinite tuples of 0 and 1 in all combinations. This set is isomorphic to all strings of 1 and 0. Suppose this set is countably infinite, so we could write

$$\begin{aligned} w_1 &= 1001101.... \\ w_2 &= 0011110101.... \\ w_3 &= 1111001110100000111 0011..... \\ w_4 &= ..... \\ etc \end{aligned} \tag{13.29}$$

This is much like the strings of section 9.3, except the strings above do not represent numbers - they are just strings of 1 and 0 - and hence termination by all 1 is permissible. The same cardinality argument used in section 9.3 is valid for this case, and therefore the set, equation 13.28 is uncountable.

What is interesting about this result is that the cardinality of a cartesian product of a **finite** number of sets $\{0,1\}$ is finite. This is true for any finite number of factors in the product. Yet the cardinality of a countably infinite product is **uncountable** and not countably infinite. That is, in passing from finite to countably infinite product, for the cardinality of the product, $\aleph_0$ is bypassed and cardinality jumps right to $\aleph_1$.

This result generalizes as follows.

Theorem 13.9

The countably infinite cartesian product of any finite set is uncountable. That is, if S is a finite set, then

$$\prod_{i \in \mathbb{Z}_+} S = \prod_{i=1}^{\infty} S \cong \mathbb{R} \tag{13.30}$$

We use the same argument. Suppose $S = \{a, b, c\}$, abstract entities. Then the cartesian product is isomorphic to all strings of the form

$$\begin{aligned} w_1 &= abaacbaabc.... \\ w_2 &= babbcaacccb.... \\ w_3 &= aaccccccaababbc..... \\ w_4 &= ..... \\ &etc \end{aligned} \tag{13.31}$$

Again we create a string $q = q_1q_2q_3...$ which is not in the above list. For w_1, if the first slot is 'a' or 'b', set $q_1 = c$ otherwise set $q_1 = a$. Similarly for the second slot of w_2, the third slot of w_3 and so forth. The resulting string q differs from all w_i in slot i and therefore the product is uncountable.

14 symbolic logic and set theory

This section can be skipped on a first reading.

Set theory is a part of logic. Logic is a large and complex subject, and here we consider only a part of it - just enough to prove several interesting theorems and to give a more general discussion of set definition than was given in section 5.2.

14.1 propositions, expressions and the = symbol

A **proposition** is an expression whose only values are True (T) and False (F). Propositions can be formed in several ways, and those of interest will be illustrated by example.

First we consider a **logical mathematical expressions**, which is a kind of proposition. A logical mathematical expression is a mathematical expression, or **statement**, whose value is T or F. Here are some examples of True propositions which are True logical mathematical expressions (see the following box).

Let $A = \{1, 2, 3, 4, 5\}$, $B = \{3, 4, 5, 6, 7\}$.

$$\begin{aligned} &2 + 2 = 4 \\ &3 \in A \\ &\int_0^1 e^x dx = e - 1 \\ &5 \in A \cap B \\ &99 \in \mathbb{Z}_+ \\ &\pi \notin \mathbb{Z}_+ \end{aligned} \tag{14.1}$$

The expression $2+2$ is a mathematical expression but not a logical mathematical expression (not a statement) because its value is 4 and not T or F. Also, 4 by itself is not a proposition but is a mathematical expression. But $2 + 2 = 4$ is a proposition (a logical mathematical expression) because it is a *true statement.* The statement as a whole has a value. Similarly, $A \cap B$ is a mathematical expression but not a proposition (not a logical mathematical expression) because its value is a set. But $A \cap B = \{3, 4, 5\}$ is a proposition which is T. It is a *true statement.*

A logical mathematical expression is a type of proposition. But *proposition* is a more general term. All the expressions above have a *value*, which in this case is T. They are propositions. This idea is explained in the following box.

Definition 14.1

The equal sign (=) is overloaded in this symbolic logic section, with an implication not used in the rest of the book.

Let A, B be some mathematical expression or proposition. ***The statement A = B has a value - T (True) or F(False).*** *The value of A = B is T when A and B have the same value. Otherwise it is F.*

In other words, ***a statement such as 2 + 2 = 4 is a proposition and has a truth value - T or F*** *- in this case T.*

Apart from this section, in the rest of this book the equal sign = is either an assignment or means that the expressions on either side of = have the same value.

For example, the proposition $\int_0^1 e^x dx = e - 1$ (equation 14.1) means that the expressions on the left side and the right side have the same numerical value, *and the statement as a whole is a proposition with a value* - in this case the value is T.

The symbols $>$, $<$, $\geq$, $\subset$, $\notin$, ... all are interpreted similarly. An logical mathematical expression such as $A \subset B$ has the value T or F depending on whether A is or is not a subset of B.

In this *symbolic logic* section, **every statement has a value - either T or F**.

So for equations 14.1, the statement $\pi \notin \mathbb{Z}_+$ has the value T because π is not a member of the set $\mathbb{Z}_+$. Similarly the statement $5 \in A \cap B$ is T because 5 is indeed in the intersection.

In many computer languages, the symbol = is an assignment, and special built-in symbols such as ==, ge, le, <, >, etc, are used to form statements which are T or F.

Here are some examples of False propositions (logical mathematical expressions which have the value F):

$$\begin{aligned} &2+2<0 \\ &20 \in A \\ &\int_0^1 e^x dx = 0 \\ &5 \notin A \cap B \\ &99 \notin \mathbb{Z}_+ \\ &\pi \in \mathbb{Z}_+ \end{aligned} \tag{14.2}$$

The value of a logical mathematical expression can be assigned to a variable, and that variable is called a proposition. We use the **assignment symbol**

$$:=$$

Let p be a **variable** which is a proposition. The assignment of the *value* of the statement $2+2=4$ to p is written

$$p := 2+2=4 \tag{14.3}$$

The value of p is T (True).

Next are expressions involving the variable p defined above, equation 14.3. For example, the following is a **logical expression**: $p = T$. This is a proposition whose value is True. So the variable p is a proposition and the logical expression $p = T$ is a proposition whose value is T because p and T have the same value (see definition 14.1). The logical expression $p = F$ is F (False) because p as previously defined is T (the left and right sides are not the same).

Consider $q := p = T$ (same p as above). The variable q is a proposition whose value is T. On the other hand, for the assignment $qq := p = F$, the variable qq is a proposition whose value is F because the statement $p = F$ has the value F.

Note that an assignment such as $p := 2+2=4$ does not have a value as a whole. It simply means that p gets the value of the statement $2+2=4$. Assignment staterments do not have a value.

> The difference between = (is equal to, is the same value as) and := (is assigned to) must be kept clearly in mind.

Here are examples of assigning logical mathematical expressions (propositions) to variables which are propositions.

$$\begin{aligned} p &:= \int_0^1 e^x dx = e - 1 \\ q &:= \int_0^1 e^x dx = 0 \\ r &:= \pi \notin \mathbb{Z}_+ \\ s &:= 5 \notin A \cap B \\ t &:= p = F \end{aligned} \tag{14.4}$$

Then p = T , q = F, r = T, s = F, t = F. The expressions p = T, q = F,..... all have value T and each are logical expressions.

> From now on, the term **expression** will mean one of: mathematical expression, logical mathematical expression or logical expression, the meaning dependent on context. The term **proposition** will mean one of: a variable whose value is T or F, a logical expression or a logical mathematical expression.

Here are sample sentences with this terminology (sets A, B above equation 14.1): "The proposition $2 + 2 = 4 \in \{T, F\}$ is T". "For the assignment $p := A \subset B$, the proposition p is F because the expression $A \subset B$ is F."

14.2 logical operators, expressions, truth tables

A **compound proposition** is an expression containing **logical operators**. A compound proposition contains one or more logical operators and one or more propositions, and is a new proposition.

The logical operators of interest here are unary and binary **operators**. An operator is really a function. A unary operator is a function of one variable, a binary operator is a function of two variables. For propositions p, q, the function is written not $F(p, q)$ but rather as a binary operator $p\ op\ q$. In the case of a unary operator, $F(p)$ is written as $op\ p$. When functions are notated as operators, the notation is called **infix notation**.

The unary operator **not** is represented by the symbol $\neg$. If proposition p is T (F), then the compound proposition $\neg p$ is F (T). A proposition q may be defined by the assignment $q := \neg p$. The proposition q is F (T) when p is T(F).

A very useful way of describing the possible values of compound propositions is with the **truth table**. Here is the truth table for $\neg$.

p	$\neg p$
T	F
F	T

Table 2: $\neg$ operator **not**

The first column lists all the possible values of *proposition* p, the second column lists the corresponding values for the proposition $\neg p$.

In using a truth table for any particular expression, one selects the row which corresponds to the truth value of the particular proposition under consideration. For example, suppose $p := 2+2=5$. Then p is F, and from table 2, row 2, when p is F, then $\neg p$ is T. In other words, $\neg(2+2=5)$ is T.

For another example, for the sets $A = \{1,2,3,4,5\}$, $B = \{3,4,5,6,7\}$, let $p := A \subset B$. Then p is F and $\neg p$ is T ($\neg(A \subset B)$ is T and $(\neg(A \subset B)) = T$ is T).

The binary operator $\wedge$ is the **and** operator. (In section 5.2 this operator was denoted &.) The expression $p \wedge q$ - read *p and q* - is T if both p and q are T, and F otherwise. Here is the truth table.

p	q	$p \wedge q$	$\neg(p \wedge q)$
T	T	T	F
F	T	F	T
T	F	F	T
F	F	F	T

Table 3: $\wedge$ operator **and**

The third column of this table shows the values of the proposition $p \wedge q$ for all possible combinations of values of propositions p and q, given in the first two columns. (This defines the symbol $\wedge$.) The fourth column lists the corresponding values of $\neg(p \wedge q)$, the values taken from table 2. The expressions $p \wedge q$ and $\neg(p \wedge q)$ are **compound propositions**.

For example, suppose $p := 2+2=4$ and $q := 3+4=22$. Then p is T, q is F, and so, from table 3, third row, $p \wedge q$ is F. This means that the compound proposition $(2+2=4) \wedge (3+4=22)$ is F. The expression $(2+2=4) \wedge (3+4=22) = F$ is T.

For another example, consider

$$(A = \{1,2,3,4,5\},\ B = \{3,4,5,6,7\} \tag{14.5}$$

and the proposition

$$q := (|A| = 5) \wedge (|B| = 8) \tag{14.6}$$

The expression $|A| = 5$ is T and the expression $|B| = 8$ is F. Therefore the value of the proposition q is F. That is, q = F is T, and the proposition r defined as $r := (q = F)$ has value T.

A third operator is **or**, with symbol $\vee$. The expression $p \vee q$ reads *p or q*. The value of $p \vee q$ is T if p or q or both are T and F otherwise. This is an **inclusive or**. The truth table is below.

p	q	$p \vee q$	$(p \vee q) \wedge p$	$p \vee (p \wedge q)$
T	T	T	T	T
F	T	T	F	F
T	F	T	T	T
F	F	F	F	F

Table 4: $\vee$ operator **or**

This table has two additional columns with compound propositions. We see that the truth values of columns 1, 4 and 5 are identical.

Two proposition are equal if the corresponding columns in the truth table are identical. Otherwise they are not equal.

The propositions are also said to be **equivalent**.

The propositions p, $(p \vee q) \wedge p$, $p \vee (p \wedge q)$ are equivalent. So for example, for any values of p and q, the expression $p = (p \vee q) \wedge p$ is T.

Now we consider the **conditional implies** binary operator $\rightarrow$, also referred to as the **conditional** operator. (The term *implies*, used by itself, is reserved for something else.) Here is the truth table.

p	q	$p \rightarrow q$
T	T	T
F	T	T
T	F	F
F	F	T

Table 5: $\rightarrow$ operator **conditional implies**

The expression $p \rightarrow q$ is frequently interpreted as **if p then q**, or **the truth value of q is conditional on the truth value of p** or **the truth value of q is determined by the truth value of p**.

Unfortunately, the interpretation is not entirely straightforward because, first, the truth table may be a little confusing and, second, what does it mean that there exists a truth table for conditional implies?

A truth table gives values for the entire expression $p \to q$, so the logical expression $p \to q$ **as a whole** has a truth value which is determined by the combined values of p and q. That is, **the value of q is not deduced from p. Rather, p and q have values (determined by a particular situation - see "models" below), and the value of $p \to q$ is then computed from the table.**

The interpretation *if p then q* could mean something like this: If p is T and if $p \to q$ is (somehow) found to be T, then we may deduce that q is T, since only row 1 of the above table applies.

One aspect of this table which may be a little confusing involves the second and fourth rows. If p is F, then no matter whether q is F or T, the expression $p \to q$ is T. If we insist on an *if p then q* interpretation, then expanding the paragraph above, if p is F and $p \to q$ is (somehow) found to be T, then we may deduce that q can be both T (row 2) **and** F (row 4) at the same time (an unhappy conclusion). In other words, we may deduce any proposition q we like. We may deduce that both propositions $2+2=4$ and $2+2=5$ make $p \to q$ have value T.

The following illustrates the above statement.

$$\begin{aligned} \pi \in \mathbb{Z}_+ &\to 2+2=4 \\ \pi \in \mathbb{Z}_+ &\to 2+2=5 \end{aligned} \tag{14.7}$$

The left side is F for both statements, so only rows 2 and 4 of table 5 are applicable, and therefore each statement as a whole has value T. (The right side is T for the first statement and F for the second). Alternatively, given a False p, and assuming $p \to q$ is T, we deduce a contradiction (a false p implies both a True q and and False q). This is generally taken to mean that, from a False assumption, we may deduce anything at all.

The third row of table 5 is generally taken to mean that a T proposition should not imply a proposition which is F.

It is left as an exercise to show the following equivalences:

$$p \to q = (\neg p) \vee q = \neg(p \wedge \neg q) = (p \wedge q) \vee (\neg p) \tag{14.8}$$

Note that the conditional, $\to$, is not an independent operator. It can be written in terms of $\wedge$, $\vee$ and $\neg$. Therefore, its interpretation as *implies* may be conditional.

The **biconditional** operator $\leftrightarrow$ is defined as follows:

$$p \leftrightarrow q = (p \to q) \wedge (q \to p) \tag{14.9}$$

Exercise: draw the truth table for the biconditional.

14.3 models and truth tables

We have encountered three types of expressions which are propositions. The list below illustrates the three types, where p, q, r, s, t, u, v are propositions.

(1). $s := (2 + 2 = 4),\ \ t := (A \cap B = \{3, 4, 5\})$
(2). $r := p \wedge q$
(3). $u := v \vee (2 + 2 = 4)$

where parens are for clarity. Expressions (1) are mathematical expressions. (2) is a purely logical expression. (3) is a combination.

The question is: **What kind of truth tables are relevant for these expressions?**

Neither expression of (1) needs a truth table. The T, F values of the mathematical expressions are determined by the usual mathematics or set theory, and so s and t are both T. There are no alternative values. The **valuations** producing the T, F values of the right side of each assignment are made **outside** the domain of symbolic logic. The valuations are carried out by a **model**. In this case, the **models** are arithmetic and set theory.

In the case of (2), the truth value of r is determined by the truth values of p and q. We don't have any information about these values, because we don't know what p and q represent. All we know is that they are propositions, so their only possible values are T and F. In any particular instance, there could be a **model (or function)** M which values both p and q as T ($M(p) = M(q) = T$). Then the value of r is determined by the first row of table 3 for the $\wedge$ operator. These valuations of the propositions p, q are carried out by *model1*. A different model to value p and q might result in p = T and q = F. In this case the value of r is determined by the third row of table 3. Let *model3* be the function which does this valuation. It is a different model than model1.

We can include the models as part of the truth table if we like, so that table 3 might look something like

model	p	q	$p \wedge q$
model1	T	T	T
model2	F	T	F
model3	T	F	F
model4	F	F	F

Table 6: $\wedge$ operator with models

In other words, **each row of a truth table corresponds to a different model** to calculate p and q. The models are different because they give unique values of the p,q pair. In any particular situation, one of the models is expected to be applicable. The truth table illustrates the effects of all possible models. There could be a model644 which has the same T, F effect as model3, so row three would correspond to model644 as well as to model3.

For expression (3), the value of $2 + 2 = 4$ (which is T) is determined by the arithmetic model. As far as we are concerned here, there is no other model. On the other hand, the value of v is not necessarily known and therefore the truth table might look something like below:

model	v	$2+2=4$	$v \vee (2+2=4)$
modelA	T	T	T
modelB	F	T	T

Table 7: $\vee$ operator with models

The two models, modelA and modelB, value only v and not the arithmetic expression. The model for that expression is implicit.

14.4 theorems: implies $\Rightarrow$ and conditional implies $\rightarrow$

A mathematical structure is made up of theorems which are built on a foundation of assumptions (axioms) and other theorems. A theorem normally has the form

$$p \Rightarrow q \tag{14.10}$$

where p and q are propositions. This reads: **p implies q**, or, alternatively: Given p, then q follows; Given p we deduce q; Given that p is True, then q is True. Of course, the fact that q is T, given that p is T, must be proven.

For example: Given sets X, Y, the following is a theorem:

$$(X \cap Y = Y) \Rightarrow Y \subseteq X \tag{14.11}$$

(To prove theorem 14.11, assume the left side is T but the right side is F. Then there exists a $y \in Y$ such that $y \notin X$, in which case $X \cap Y \neq Y$, a contradiction.)

This theorem can be notated as follows:

$$\begin{aligned} &p := X \cap Y = Y \\ &q := Y \subseteq X \\ &p \Rightarrow q \end{aligned} \tag{14.12}$$

The symbol $\Rightarrow$ means **implies** and **it is not a symbol of symbolic logic**. The statement $p \Rightarrow q$ is **not a statement (or expression) of symbolic logic**. Rather, it is a **meta-statement** - a statement **about** expressions of symbolic logic. The mathematical meaning of equation 14.11 or the last of equations 14.12 involves models.

> Meta-statements in combination with models give meaning to the otherwise formal and uninterpretable statements of pure symbolic logic.

Meta-statements **connect** logical expressions to each other in a way that makes sense for some purpose.

For example, consider equations 14.11 and 14.12. The $\Rightarrow$ symbol works as follows. The *model* to evaluate the left side is set-theory. We are only interested in models where the sets X, Y are such that $X \cap Y = Y$ has value T. (That is, we consider only models where p = T.) When this is the case, then the **same exact models** are used to evaluate the right side, $Y \subseteq X$. If this evaluates to T **by all of the exact same models that give p = T**, we say that the theorem, equation 14.11, is **valid** or **established**.

Speaking generally now about any theorem of the form $p \Rightarrow q$ (equation 14.10), **if p is true according to one or more models, and if q is T from the same models, then the theorem is deemed valid**. There are serious repercussions, because then we (as mathematicians or scientists) will go further, perhaps, to say that the theorem contributes to an existing mathematical structure, or that the theorem implies the result of some observation (experiment) on the real world. This is an interpretation of the interpretation.

It is *not* meaningful to say that a theorem has the value T, or is T or is True. We could say the theorem is "true", but we use (or at least imply) quotation marks, because "true" means the same as "valid". A theorem equation 14.10 does not have a truth value. Even though p and q individually are T or F (have truth values), the statement $p \Rightarrow q$ does NOT have a truth value. It is a meta-statement, it connects propositions. It is a statement about propositions. It is a theorem based upon particular models.

We now relate *conditional implies* $\rightarrow$ to *implies* $\Rightarrow$.

The following definition gives the meaning of a theorem being valid.

Definition 14.2

If M is a model and p is a proposition, then $M(p)$ is the valuation of the proposition based on the model. The quantity $M(p)$ can have one of only two values - T or F.

Given a proposition p, let M_k be models, where there exists an index set K such that $M_k(p) = T$ for all $k \in K$.

The index set indexes all possible models where $M_k(p) = T$.

For a theorem $p \Rightarrow q$, assume $M_k(p) = T$ for all $k \in K$ (this defines the set K). Then

$$\text{If } M_k(q) = T \text{ for all } k \in K, \text{ then } p \Rightarrow q \text{ is valid.}$$

Now consider the table:

M_k	p	q	$p \to q$
M_1	T	T	T
M_2	T	T	T
M_3	T	T	T
....	T	T	T
....	....	....	....

Table 8: models and conditional

where column 1 lists all the models M_k where $k \in K$ (ie, all models where p is T and also q is T). From table 5, the expression $p \to q$ is T for all rows. In other words

Theorem 14.3

Given p, q and models M_k as above (where $M_k(p)$ is T), then

$$p \Rightarrow q \text{ iff } p \to q = T$$

Referring to table 5, if p is T and $p \Rightarrow q$ is assumed valid, then q is T (definition 14.2), hence p and q are T and so $p \to q$ is T. If p is T and $p \to q$ is T then the only possible value for q is T , and therefore $p \Rightarrow q$.

14.5 logical quantifiers - predicate calculus

To define sets in a general manner we must enhance our definition of propositions to include functions. A **predicate** is a proposition whose truth value depends on the values of one or more variables. In other words, **a predicate is a function** which can take on only one of two values: T or F. If P is a predicate and S is some set on which P is defined, then

$$P : S \to \{T, F\} \tag{14.13}$$

For example, let x be a real number, $x \in \mathbb{R}$. Then we may define a predicate P(x) ($P : \mathbb{R} \to \{T, F\}$) as

$$P(x) := 2 + x = 4 \tag{14.14}$$

P(x) is a proposition whose value (T or F) depends on the variable x. We can have, for example (p, q, r are propositions):

$$p := P(3), \quad q := P(6), \quad r := P(2) \tag{14.15}$$

Then the values of p, q are F while the value of r is T. In this case, P(x) is true only for one particular value of x, and is F otherwise. Note that in this particular case P is surjective but not injective.

Consider predicates $R, W : \mathbb{R} \to \{T, F\}$. The predicate $W(x) := x^2 > 0$ is true for all real x except $x = 0$ while $R(x) := x^2 < 0$ is F for all x. W is surjective while R is not surjective, and neither are injective.

This brings us to the idea of a **quantifier**. For equation 14.14, we can say: **There exists some real number x for which P(x) is T.** This sentence is equivalent to the mathematical expression

$$\exists x \in \mathbb{R} \mid (P(x) = T) \tag{14.16}$$

which reads: There exists ($\exists$) some $x \in \mathbb{R}$ (at least one) such that ($|$) the predicate $P(x)$ is T. The expression 14.16 is a logical expression of predicate calculus and so has the value T or F. We can assign it to a variable p:

$$p := \exists x \in \mathbb{R} \mid (P(x) = T) \tag{14.17}$$

The variable p is a proposition. The value of p in this particular case is T, because there does exist an x for which P(x) is T (equation 14.14). The symbol $\exists$ is called an **existential quantifier**. So the following expression is T: $(\exists x \in \mathbb{R} | (P(x) = T)) = T$. It may be amusing that $(\exists x \in \mathbb{R} | (P(x) = T)) \wedge (\exists x \in \mathbb{R} | (P(x) = F))$ is T.

We may write the expression 14.17 in a shorthand way. Defining variable x to be a real number we can write 14.17 as

$$p := \exists x \mid P(x) \tag{14.18}$$

where P(x) standing alone is interpreted as $P(x) = T$. In practice, the set from which x is drawn must be defined explicitly or implicitly. In equation 14.18, it is implicit that x is drawn from the set $\mathbb{R}$. In equation 14.17, that set is explicit.

Definition 14.4

Given non-empty set S and predicate P(s) which is defined for all $s \in S$, the proposition

$$\exists s \in S \mid P(s) \tag{14.19}$$

is T if P(s) is T for at least one $s \in S$, and is F otherwise.

Equation 14.19 is equivalent to $\exists s \in S \mid (P(s) = T)$. The case of S as the empty set is discussed later.

In these quantified expressions, the predicate P(s) is any logical expression involving the variable s. So the expressions may include logical operators. For example, say $P(x) := 2 + x = 4$

$$q := \exists x \in \mathbb{R} \mid (P(x) \wedge (x < 0)) \tag{14.20}$$

states that q is T if there is at least one real number x such that P(x) is T and x $<$ 0. Since such is not the case, then $q = F$. On the other hand

$$r := \exists x \in \mathbb{R} \mid (P(x) \vee (x < 0)) \tag{14.21}$$

is T.

Let A, B be the sets of equation 14.5 ($A = \{1, 2, 3, 4, 5\},\ \ B = \{3, 4, 5, 6, 7\}$), the proposition

$$q := \exists x \in A \mid x \in B \tag{14.22}$$

reads: There exists at least one x in A which is also in B. Therefore q = T. Note that the above proposition is equivalent to a logical expression which does not involve a quantifier:

$$(\exists x \in A \mid x \in B) = (A \cap B \neq \emptyset) \tag{14.23}$$

where the parens are for clarity. Both expressions are T. The following proposition is F for A, B as above:

$$\exists x \in (A \cap B) \mid (x > 5) \vee (x < 0) \tag{14.24}$$

If A, B are arbitrary sets, then

$$(\ (\exists x \in A \mid x \in B) = (A \cap B \neq \emptyset)\ ) = T$$

because the propositions on either side of the first = are both T or both F together. We also have the following:

$$\begin{aligned} q &:= \exists x \in \mathbb{R} \mid x^2 > 0 \\ r &:= \exists x \in \mathbb{R} \mid x^2 < 0 \end{aligned} \tag{14.25}$$

q is T and r is F.

Now consider the predicate $Q(x) := x^2 \geq 0$. The statement: Q(x) is T for all $x \in \mathbb{R}$ is

$$w := \forall x \in \mathbb{R} \mid (Q(x) = T) \tag{14.26}$$

which reads: For all (∀) real numbers x the predicate Q(x) is T. The symbol ∀ is called the **universal quantifier**. The value of the logical expression on the right side of equation 14.26 is assigned to w, so w is T, and the logical statement $w = T$ is T. In the expression 14.26 above, the vertical bar "|" is a delimiter and is "silent".

On the other hand, the value of the expression $\forall x \in \mathbb{R} \mid (x^2 > 0)$ is F because 0^2 is not positive.

For sets A, B, equation 14.5, the expression

$$\forall x \in A \mid x \in B \tag{14.27}$$

is F (for example, $1 \in A$ but $1 \notin B$).

Definition 14.5

Given non-empty set S and predicate P(s) which is defined for all $s \in S$, the proposition

$$\forall s \in S \mid P(s) \tag{14.28}$$

is T if P(s) is T for for every $s \in S$, and is F otherwise.

Proposition equation 14.28 is equivalent to $\forall s \in S \mid (P(s) = T)$ The case of S as the empty set is discussed later.

Any proposition can be negated, so we consider the negation of propositions involving quantifiers. Let

$$p := \exists x \in \mathbb{R} \mid (2 + x = 4) \tag{14.29}$$

The value of p is T. The value of $\neg p$ is F. The expression

$$\neg(\exists x \in \mathbb{R} \mid 2 + x = 4) \tag{14.30}$$

means that there does not exist in $\mathbb{R}$ any number x such that $2 + x = 4$. In other words, this means that **for all** $x \in \mathbb{R}$, $2 + x = 4$ is F. That is

$$\neg p = \neg(\exists x \in \mathbb{R} \mid 2 + x = 4) = \forall x \in \mathbb{R} \mid \neg(2 + x = 4) \tag{14.31}$$

and the value of $\neg p$ is F. In general, then

Definition 14.6

Given non-empty set S and predicate P(s) which is defined for all $s \in S$,

$$\neg(\exists s \in S \mid P(s)) = \forall s \in S \mid \neg P(s) = \forall s \in S \mid \neg(P(s) = T) \tag{14.32}$$

If there is no s in S such that P(s) is T, then for all s in S, P(s) is F ($\neg(P(s) = T) = T$). The case of S as the empty set is discussed later.

Since for any proposition p it must be the case that $\neg\neg p = p$, the above definition gives

Definition 14.7

Given set S and predicate P(s) which is defined for all $s \in S$,

$$\neg(\forall s \in S \mid P(s)) = \exists s \in S \mid \neg P(s) \tag{14.33}$$

In other words, if it is not the case that P(s) is T for all s in S, then there exists some s in S (at least one) such that P(s) is F. Again, the case of S as the empty set is discussed later.

14.6 creating sets using predicates

Section 5.2 presented a notation for creating sets. Let S be a set, let $F : S \rightarrow F(S)$ be a *bijective* function and let Q(s) be a predicate defined for all $s \in S$.($Q : S \rightarrow \{T, F\}$.) Then the following defines a set X in the notation of section 5.2

$$X = \{F(s) \,|\, \forall s \in S \,\&\, Q(s)\} \subseteq F(S) \tag{14.34}$$

The set X is a subset of the image of S under F. Only those values of F(s) are in X for which Q(s) is T. If Q(s) is T for all $s \in S$, then $X = F(S)$ because F is assumed to be bijective. *The symbol & in this expression is supposed to be the same as $\wedge$ of symbolic logic.*

Equation 14.34 is **not** a statement of symbolic logic. For one thing, the string $\forall s \in S$ does not have a truth value. Therefore the string $\forall s \in S \,\&\, ...$ is not really defined. Rather, **equation 14.34 is a shorthand for a union**, as discussed in section 5.2. A more formal expression for the set X can be written as follows. Define a function $\{.\,|\,.\}$ as

$$\{F(s)\,|\,Q(s)\} = \begin{pmatrix} \emptyset & Q(s) = F \\ \{F(s)\} & Q(s) = T \end{pmatrix} \tag{14.35}$$

where $s \in S$. The function $\{.\,|\,.\}$ always returns a set - either the empty set or the singleton $\{F(s)\}$. Then the expression 14.34 is shorthand for

$$X = \emptyset \cup \{F(s)\,|\,Q(s)\} \cup \{F(s')\,|\,Q(s')\} \cup \{F(s'')\,|\,Q(s'')\} \cup .... \tag{14.36}$$

where $s, s', s'', ... \in S$. **Notation 14.36 is such that, if S is empty, then** $X = \emptyset$. If S is not empty, but if Q(s) is F for all s in S, then also $X = \emptyset$.

Suppose S is not empty. Then we can directly link set creation with the universal quantifier of predicate calculus

$$X = F(S) \iff \forall s \in S\,|\,Q(s) \tag{14.37}$$

where $\iff$ means *implies in both directions*. In the next section we use this equation to determine the *edge conditions*, when universal and existential quantifiers reference empty sets.

14.7 predicates and the empty set

We now return to the definitions 14.4 and 14.5 when the set S of those definitions is the empty set. The goal is to find end conditions for these expressions, if possible. Consider the following:

$$\begin{aligned} &p := \forall s \in S\,|\,Q(s) \\ &X = \{s\,|\,\forall s \in S\,\&\,p\} \\ &X = S \iff p = T \end{aligned} \tag{14.38}$$

for **any** predicate Q whatsoever and S nonempty. The middle expression is shorthand for a union.

Now suppose that S is the empty set. While strictly speaking the first and third of 14.38 are not meaningful if S is empty, in developing some theory, one might come up with a set S which may be empty or not be empty - one may not know in advance, and either could happen. So we should try to do something for S empty which is at least logically consistent with what happens if S is not empty, and which makes some kind of (peculiar) sense.

It has been established that the set X in equation 14.38 must be the empty set when S is empty, regardless of Q (however defined). Therefore, if S is empty, the third equation of 14.38 requires that p be equal to T, and vice versa. The conclusion is:

For any predicate Q(s) defined on some non-empty set S,

$$(\forall s \in \emptyset\,|\,Q(s)) = T \tag{14.39}$$

Equation 14.39 is really a consistency requirement - an edge condition. It does not actually mean anything because the predicate Q is never evaluated. It is valid for any Q (even constant Q, such as Q(s)=T) and so is content-free. (We will see that it should not be used.)

Note that, if we only considered the first two expresssions of equation 14.38, then we could also conclude that $\forall s \in \emptyset \,|\, Q(s) = F$. (Exercise: show this) However, if we insist that the third expression is also valid (which is important in defining an edge condition), then equation 14.39 is the only conclusion. The edge condition should be an extrapolation from non-empty S to empty S.

Then applying definition 14.7 to equation 14.39 we find (letting $H = \neg Q$)

For any predicate H(s) defined on some non-empty set S,

$$\exists s \in \emptyset \,|\, H(s) = F \tag{14.40}$$

the related consistency condition (if Q is arbitrary, so is H).

Unfortunately these consistency conditions are not consistent. If S were not empty, then if some $\forall s \in S|$..... were True, the related $\exists s \in S|$.... should also be True. If the edge conditions are extrapolations, then equation 14.40 should be T and not F.

Therefore

The edge conditions 14.39 and 14.40 should not be trusted in an informal approach.

Make sure to apply quantifiers to non-empty sets.

14.8 intersection when an index set is empty

Consider an indexed set of sets

$$S = \{S_a, S_b, S_c, ...\} \tag{14.41}$$

and some set I of indices which indexes all the sets of S. In this section, assume S is not empty.

Define X as the union of **all** sets in S,

$$X = \bigcup S \tag{14.42}$$

Let sets S and X be fixed once and for all. If I is not empty, then also assume that if $w \in I$ then $S_w \in S$ (but not necessarily the inverse). Then

$$\bigcup_{w \in I} S_w \subseteq X = \bigcup S \tag{14.43}$$

We want to compute the sets

$$\begin{aligned} U &= \bigcup_{w \in I} S_w \\ Y &= \bigcap_{w \in I} S_w \end{aligned} \tag{14.44}$$

when I is the empty set. From the previous sections, it is clear that, if $I = \emptyset$ then $U = \emptyset$ (definition 4.2). It is also clear that that $Y \subseteq X$ when I is not empty, but what is Y when $I = \emptyset$?

An alternative specification of Y is (see definition 4.6)

$$Y = \{x \,|\, \forall x \in X \wedge (\forall w \in I \,|\, x \in S_w)\,\} \tag{14.45}$$

Now suppose I is empty. This does not modify S or X or any S_w. But if I is empty, equation 14.39 gives the following: $(\forall w \in \emptyset \,|\, x \in S_w) = T$ for any particular x in X. Therefore, when I is empty

$$Y = X \tag{14.46}$$

To summarize:

Theorem 14.8

Given a nonempty set of sets $S = \{S_a.S_b, S_c, ...\}$. *Let* I *be a set of indices such that, if* I *is not empty and if* $w \in I$, *then* $S_w \in S$ *(but not necessarily the inverse). Let* $X = \bigcup S$. *Then*

$$\begin{aligned} \bigcup_{w \in \emptyset} S_w &= \emptyset \\ \bigcap_{w \in \emptyset} S_w &= X \end{aligned} \tag{14.47}$$

This is a peculiar result. It is expected that an intersection of sets should be a subset of or equal to a union of those sets. However, for an empty index set, we have that the union is a subset of the intersection, just the opposite of expectations. While we have seen that the first expression has some reasonable interpretation, the second equation appears to be meaningless, and in fact might be a problem when comparing the sizes of sets.

Index sets are meant to be a useful notation. The above result indicates that

Unions and intersections over empty index sets are to be explicitly avoided.

14.8.1 intersection of an empty set of sets

Let S be a set of sets as above, but without an index set. For example, $S = \{A, B, C, ...\}$ where A, B, C are sets. Let $X = \bigcup S$ as above.Then the intersection of all sets in S can be written

$$Y = \bigcap_{R \in S} R = X \cap A \cap B \cap C... \tag{14.48}$$

from which we get the analog of equation 14.45 (different Y)

$$Y = \{x \,|\, x \in X \wedge (\forall R \in S \,|\, x \in R)\} \tag{14.49}$$

By the same argument as section 14.8 above, we see that if S is the empty set, Y = X. Therefore we must conclude, in analogy with the case of an empty index set, that

Unions and intersection over an empty set of sets are to be explicitly avoided.

Of course a union over a non-empty set of empty sets is well-defined.

15 paradoxes and set specification

Definition 15.1

In order to create a set S from elements of a nonempty set X (so that $S \subseteq X$) , we must be able to determine ***unambiguously*** *whether any element of X is or is not a member of S. If this cannot be done, then S is not a valid set. Any such situation is called a* ***paradox****. In the context of this book, a paradox is a set specification in which membership of some element cannot be resolved.*

There are many other paradoxes which are different from this definition, but it is sufficient for purposes of this book. For as long list of paradoxes see *en.wikipedia.org/wiki/List_of_paradoxes*.

Here is an example - the **library paradox**. Consider set X of all books in a library. (We can call X a library.) Define the subset G as those elements of X which do not refer to themselves. That is, if a library book title is ABCD, then if the book is to be a member of G the word ABCD may not appear anywhere in the book apart from the title page and the cover.

Now consider a book b which is a catalog of **all** the books in G. The book b is in X. The question is: is b in G? Since b is a catalog of all books in G, then if it is in G, it must refer to itself (since it is a catalog of **all** books in G). Therefore it cannot be in G, a contradiction. On the other hand, if it is not in G, then it does not refer to itself (since it lists only books in G) and therefore is suitable for being in G. So it must be in G, in which case it must list itself, a contradiction. It cannot be resolved whether or not the catalog is in G. This is a paradox, and G is not a set.

This paradox is specially formulated to **be** a paradox. There is some little twist which is weird but which makes it a paradox.

15.1 Russell's paradox and set specification

The paradox discussed above is of course artificial and can be made non-paradoxical by not calling the library catalog a book. Russell's paradox is a paradox which arises naturally in set theory and places a significant constraint on what kinds of mathematical objects can be legitimately called sets and which, therefore, can be manipulated by operations of set theory. It is not specially formulated to be a paradox, and for a while it threatened the very existence of set theory as a consistent mathematical theory. One important motivation of the ZF or ZFC formalization of set theory is to make sure that Russell's paradox (and other paradoxes not discussed here) do not occur. As we will see, Russell's paradox concerns sets which are **self-referential**. That is, a set definition such as $S = \{A, B, C, ..., S\}$, so **S contains itself**. It is explained below why this is a problem.

Our specification of a set uses the following notation. Let X be a given set of sets. We can define a set A based on elements of X as

$$A = \{Q \in X \mid P(Q)\} = \{Q \mid \forall Q \in X \wedge P(Q)\} \tag{15.1}$$

where P is some predicate defined for all elements of X and the expression in the middle is simply shorthand for the expression on the right, which is shorthand for a union. (More generally, we could have written $A = \{F(Q)|Q \in X....\}$ where F is some function, but for our purposes equation 15.1 will suffice.)

Why do we require $Q \in X$? Why not just write

$$A = \{Q \mid P(Q)\} \tag{15.2}$$

where Q is unrestricted - it can be any set in the universe? If X includes every set in the universe - any set at all - then equations 15.1 and 15.2 would seem to be equivalent.

Consider such an X, which includes every possible set as members. Even if there were no objects in the universe, there is still the empty set. Then $B = \{\}$ is a set, as is $C = \{\{\}, \{\{\}\}\}$, as is $D = \{\{\}, \{\{\}\}, \{\{\}, \{\{\}\}, \}, ...\}$ and so forth. Then X would be something like

$$X = \{A, B, C, \{\{\{\{\}\}\}\}, ...\} \tag{15.3}$$

Ignore how silly this is. The set X is itself a set and therefore X must be a member of X. Therefore

$$X = \{X, A, B, C, \{\{\{\{\}\}\}\}, ...\} \tag{15.4}$$

In other words, **X is a member of itself**. This is where a problem lies - the unrestricted definition of X. In this case, X refers to itself. The set X is **self-referential**.

Suppose in a world of objects that there are other sets which are members of themselves. Let M be the set of all sets in the universe which are members of themselves.

$$M = \{A \mid A \in A\} \tag{15.5}$$

Note that A is not restricted. It can be any set at all. M may or may not be a member of itself. Now define the set R as the set of all sets which are not members of themselves.

$$R = \{A \mid \neg(A \in A)\} \tag{15.6}$$

Since M and R are totally inclusive, then R, as a set in the universe of sets, must either be in one of these sets or the other. That is, it is either a member of itself or it is not a member of itself. That is

$$R \in M \cup R \tag{15.7}$$

But because R and M are mutually exclusive, we must also have

$$R \cap M = \emptyset \tag{15.8}$$

Suppose $R \in M$. In this case, $R \cap M \neq \emptyset$, which contradicts 15.8. Therefore $R \notin M$. However, because of 15.7, if $R \notin M$ then we must have $R \in R$. But this contradicts 15.6,which states that R is not a member of itself.

So we have a paradox. If R is in M, it is not in M. If R is not in M, it must be in M. We cannot decide whether R is or is not in M, and therefore M is not a (valid) set.

Backing up, assume M is indeed a valid set. We know from 15.6 that $R \notin R$. Therefore R must be in M (equation 15.7). But R cannot be in M (15.8), so it must be in R (15.7). Therefore, if R is not a member of itself, it is a member of itself. So **R cannot be a set** either.

No set which is a member of itself is a valid set. This restriction is relevant for sets of ordinal numbers, as will be seen.

Since this book describes an informal approach to set theory, creating a set based on a well-defined set might be one way of trying to avoid self-reference. The non-existence of Russell's paradox (or some other paradox) for any given set means that the definition of set creation should, at the very least, not include self-reference. While this procedure cannot guarantee the created set is valid, in practical cases it usually is sufficient. But without a formalism, there is no guarantee.

16 binary relation

It is useful here to introduce the general concept of a **relation** between two sets. Equivalence relations and maps make use of relations, as does the cartesian product of two sets.The next section - order relations - is usefully discussed referring to the general concept of relation.

Definition 16.1

Given two nonempty sets A, B, a ***binary relation*** *is any subset of* $A \times B$ *(including* $A \times A$ *if* $B = A$*).*

In other words, a binary relation R is a **set** *of ordered pairs (x,y). A relation may be notated using infix notation, so that the ordered pair (x,y) may be written xRy. A symbol is frequently used instead of R, such as* $\sim$ *or* $>$*.*

In this book, if the term relation is used by itself, it means binary relation.

Note that a given binary relation between elements of sets A, B is *usually* a **proper subset** of $A \times B$, where the relation pairs elements which relate to each other in some way. We can think of a relation as isomorphic to the set of rows of two columns of a database table. For example, if A is names and B is addresses, then a relation is the set of ordered pairs (name, address). Obviously not every address can be paired with the same name, so the ordered pairs of the relation specify something useful and the cardinality of the relation, in this example, is considerably smaller than the cardinality of $A \times B$. However, there could be several ordered pairs with the same name and different addresses and several ordered pairs with different names and the same address.

As another example, one could pair a number with its double, such as (1,2), (2,4), (3, 6),.... For another example, A could be integrands and B the integrals of integrands from 0 to 1. Exercise: Think of a few examples which might be considered unusual. For example, (paint, drying time of the paint).

17 power set

The power set may be used to define an infinite hierarchy of infinities, as discussed later. It is useful to introduce the power set here, as it provides an example of a certain type of ordering in the discussion of order relations in the next section.

Definition 17.1

A power set of a set S, notated $\mathcal{P}(S)$, is a collection of all subsets of S, including $\emptyset$. S may be finite or infinite. $\mathcal{P}(S)$ is a set of sets.

$\mathcal{P}(\emptyset) = \{\emptyset\}$.

A power set is a set of sets. For example, $S = \{a, b, c\}$.

$$\mathcal{P}(S) = \{\emptyset, \{a\}, \{b\}, \{c\}, \{a, b\}, \{a, c\}, \{b, c\}, \{a, b, c\}\} \tag{17.1}$$

If the set S is finite, then the cardinality of the power set is given by

$$|\,\mathcal{P}(S)\,| = 2^{|S|} \tag{17.2}$$

Exercise: derive the above expression.

18 order relations

18.1 definition

An order relation on a set describes some ordering of the elements of the set. For example, an ordering of numbers indicates which numbers are larger than other numbers. An ordering of items of a list indicates which items precede or succeed others. An ordering of a set of sets might indicate which sets are subsets of other sets. A dictionary is a list of words in what is called *dictionary order*, defined later.

Four different but related order relations are described in this section, from less to more restrictive.

1. partial order
2. strict partial order
3. simple (or total) order
4. well-ordering

Here is a formal definition.

Definition 18.1

An ***order relation*** *R on set A is a subset of $A \times A$. As such, it is a binary relation, a set of ordered pairs. Note that R is a set. The notation is normally abused so that the symbol R (such as $\sim$ or $>$) is used to indicate the relation itself. Thus the ordered pair (a,b) in the relation R ($(a, b) \in R$) is also written aRb.*

It is stressed that notation is abused, so that **the order relation R is a symbol as well as a set**.

Consider the set

$$A = \{1, 2, 3\} \tag{18.1}$$

Define an ordering $\mathbf{R} = <$ such that, if a $<$ b, then $(a, b) \in R$. We would say: If aRb, then $(a, b) \in R$. Notation **aRb** (specifically a $<$ b) is **infix notation**.

In this example, the ordering is strictly less than. The order relation R $=$ $<$ on the elements of set A above is the set

$$R = \{\, (1, 2), (1, 3), (2, 3)\} \tag{18.2}$$

Note that (1,1) is not a member of R because it is not the case that 1 $<$ 1. Similarly for 2 $<$1, 3 $<$ 3, etc. Thus, while the set $A \times A$ has 9 ordered pairs, the relation '$<$' has only three ordered pairs and is a proper subset of $A \times A$.

18.2 comparable sets

It is possible that the order relation may not be applicable to all pairs of elements of a set. That is, a given element of a set may not be **comparable** to some or all the other elements. Consider the set

$$B = \{1, 2, 3, a\} \tag{18.3}$$

where 'a' is not a number. It could be an abstract element or a letter of the alphabet or an automobile. Suppose the relation '$<$' is defined as in the previous example. Without an enhancement of the meaning of $<$, it is not possible to compare an integer with the element 'a' using the symbol '$<$'. Therefore the order relation R $= <$ for set B is also given by equation 18.2. The element 'a' does not appear in any ordered pair of R. We say that the element 'a' is **not comparable** to the other elements of set B.

Consider the set

$$C = \{1, 2, 3, a, b\} \tag{18.4}$$

with a,b equal to the text strings "a" and "b". Let $x, y \in C$ and define x $<$ y as follows:

- x $<$ y if both x and y are integers and x $<$ y in the usual way,
- x $<$ y if x, y are both text strings and x is alphanumerically less than y.
- otherwise x, y are not comparable

The order relation '$<$' for set C is $<= \{(1, 2), (1, 3), (2, 3), (a, b)\}$.

Another example. Let S be a set of sets and let sets $A, B \in S$ be two elements of S. Define the order relation $\subseteq$ between any two sets A,B in S as follows: if $A \subseteq B$, then $(A, B) \in \subseteq$. If $B \subseteq A$ then $(B, A) \in \subseteq$. However, it is not necessarily the case that, for two sets, one must be a subset of the other. That is, they may not be comparable. If A, B are not comparable, so neither is a subset of the other, then neither (A,B) nor (B,A) are in $\subseteq$.

The classic example of lack of comparability among some pairs of elements of a set of sets is the order relation $\subseteq$ on a power set (which is a set of sets). Suppose $S = \{1, 2, 3, 4, 5\}$. The sets $\{1\}, \{1, 2\}, \{1, 3\}, \{1, 3, 4\}, \{3, 4, 5\}$ are examples of sets in the power set $\mathcal{P}(S)$. Let $\subseteq$ be the order relation, as defined above. Then the order relation on the power set includes the ordered pair $(\{1\}, \{1, 2\})$, because $\{1\} \subseteq \{1, 2\}$. Similarly it includes the pair $(\{1, 3\}, \{1, 3, 4\})$.However, the ordered pair $(\{1, 3\}, \{3, 4, 5\}) \notin \subseteq$ because neither set is a subset of the other.

As an exercise, write as many ordered pairs in the order relation $\subseteq$ for the above example as you have patience for.

18.3 partial order

A partial order is the least restrictive of the four order relations.

Definition 18.2

*A **partial order** $\leq$ on a set A is an order relation with the following properties. For any a, b, c in A*

1. $a \leq a$ *for every* $a \in A$ *reflexive.*
2. $a \leq b$ *and* $b \leq a$ *implies* $a = b$ *anti-symmetric*
3. $a \leq b$ *and* $b \leq c$ *implies* $a \leq c$ *transitive*

A partial order is applied only to elements which are comparable. ***An element always must be comparable to itself.*** *A given element of a partially ordered set A may be comparable to all or to some or or to none of the other elements of A (apart from itself).*

A partially ordered set is called a ***poset.***

Property (1) means that every element of A is comparable to itself.

Property (2) is equivalent to the following: If $a \neq b$ and a, b comparable, then either $a \leq b$ or $b \leq a$ but not both. And, if $a = b$, this is equivalent to property (1).Property (2) is called *anti-symmetric* because, if $a \neq b$, only one of $a \leq b$ or $b \leq a$ is true.

Based on the definition, the set of integers $\mathbb{Z}$ is partially ordered under the usual $\leq$ relation. However, since all the elements of $\mathbb{Z}$ are comparable, it also has a stronger ordering, discussed below. The same for $\mathbb{Q}$ and $\mathbb{R}$.

For any set A the power set $\mathcal{P}(A)$, can be partially ordered by the $\subseteq$ relation, as discussed above. Not all elements of the power set are comparable, which is why there is only a partial order.

There are many different kinds of ordering of integers. A different kind of partial order can be assigned to $\mathbb{Z}_+$, as follows. For $n, m \in \mathbb{Z}_+$, let $n \leq m$ when n divides m (written n | m).For example, $2 \leq 2$, $3 \leq 9$, $17 \leq 153$. However, 3 does not divide 4, so 3 and 4 are not comparable. Also, n does not divide m if $n > m$. This partial order for the set $A = \{2, 3, 4, 5, 6, 7, 8\}$ is the following set of ordered pairs:

$$\leq = \{(2,2), (2,4), (2,6), (2,8), (3,3), (3,6), (4,4), (4,8), (5,5), (6,6), (7,7), (8,8)\} \quad (18.5)$$

An **equivalence relation** $\sim$ is similar to a partial order relation $\leq$, except that instead of anti- symmetry, there is symmetry. Property (2) is:

2. If $a \sim b$ then $b \sim a$.

Note that in general not all pairs are comparable. That is, $a \sim b$ will not be true for every pair a,b. That, in fact, is the whole point of an equivalence relation.

18.4 strict partial order

A strict partial order is more restrictive than a partial order. In particular, it reverses the reflexive property, so that no element is comparable to itself.

Definition 18.3 *A* ***strict partial order*** $<$ *on a set* A *is one for which:*

1. *For no* $a \in A$ *does* $a < a$. *That is, no element is comparable to itself. Therefore, the ordered pair (a,a) is not in the order relation.*
2. *For any* $a, b \in A$, $a \neq b$, *either* $a < b$ *or* $b < a$, *but not both.*
3. $a < b$ *and* $b < c$ *implies* $a < c$.

Not all elements need be comparable in a strict partial order.

One can transform a partial order to a strict partial order by removing all ordered pairs of the form (a,a) from the order relation. In the case of the power set, the order relation is $\subset$: proper subset.

18.5 simple (total, linear) order

Definition 18.4 *A **simple** order $\leq$ on a set A has the following properties:*

1. *Every pair $a, b \in A$ are comparable.*
2. *For any $a \in A$, $a \leq a$.*
3. *For any pair $a, b \in A$, then $a \leq b$ and/or $b \leq a$, and if both hold, then $a = b$.*
4. *$a \leq b$ and $b \leq c$ implies $a \leq c$.*

*A simple order is also called **total order** or a **linear order**.*

A simple order may be defined with $<$ rather than $\leq$, in which case item (2) above is not relevant. In any particular case, whether the simple order is $<$ or $\leq$ will depend on context.

The point is that every pair of elements in A are comparable.

For the second property, if a = b, then a, b are variables which stand for the same thing. Only one copy can be in the set. But it could be that, in some proof of a theorem, one finds that what is being called two elements of a set are really the same. A common proof finds that both $a \leq b$ and $b \leq a$ hold, which proves that $a = b$.

A simple order looks like a strict partial order except all pairs of elements in the set must be mutually comparable. The power set cannot have a simple order under the $\subseteq$ order relation because not all elements are comparable. The sets $\mathbb{Z}, \mathbb{Q}, \mathbb{R}$ are all simply (totally, linearly) ordered with the usual meaning for $\leq$. The **dictionary order** is an example of a simple order under $<$ on the words of a dictionary, where any two words are unequal and comparable. We illustrate with a few examples. We can order words using the order of letters of the alphabet: noLetter $<$ a $<$ b $<$ c $<$ ... $<$ z. Then art $<$ bed, ordered on the first letter. But arm $<$ artistic ordered on the third letter because m $<$ t and arm $<$ armature because noLetter after m (of arm) is less than the 2nd 'a' of armature. Then viola $<$ violas $<$ violin.

The same principle applies to the ordering of numbers. 0.13428762 $<$ 0.1342974521113, ordered on the fifth digit.

We can be a bit more formal. Consider two words $x = x_1x_2x_3...x_N$, $y = y_1y_2y_3...y_M$ where x_i, y_j are letters. Then:

1. if $x_1 < y_1$ then x < y. If $x_1 > y_1$, then y < x.
2. Now $x_1 = y_1$. If N = 1, M > 1 then x < y. If M = 1, N > 1 then y < x
3. Otherwise repeat with words $x \to x_2x_3...x_{N-1}$, $y \to y_2y_3...y_{M-1}$

18.6 well-ordering

The concept of well-ordering is important in defining ordinal numbers. In addition, there are a number of surprising results involving well-ordering. Since proving some of these requires a more formal approach, we will just mention several results.

Definition 18.5

*A **well-ordered set** is a totally-ordered set for which every subset (including the set itself) has a minimum element.*

Any totally ordered finite set is well-ordered under any ordering, so well-ordering normally refers to infinite sets with particular ordering. An infinite set can be well-ordered under one order relation and not well-ordered under another.

For example, the set $\mathbb{N}$, natural numbers (with or without 0 included) is well-ordered under the usual total order $<$, while the set $\mathbb{Z}$ is not well-ordered under $<$, because there is no minimum element and so not every subset has a minimum element. (For example, the subset of integers less than 0 is not well-ordered.) The words in a dictionary are well-ordered under the dictionary ordering. The set $\{\{1\}, \{1,2\}, \{1,2,3\}, ...\}$ is well-ordered under $\subseteq$. (Exercise: show this.)

No open subset (a,b), a < b, of $\mathbb{R}$ is well-ordered, as there is no minimum element. No closed interval [x,y], x < y, of $\mathbb{R}$ is well ordered because an open subset (a,b), x < a < b < y, does not have a minimum element. Therefore $\mathbb{R}$ is not well-ordered under $<$, and neither is $\mathbb{Q}$. The power set of any set under $\subset$ is not well-ordered because it is not totally ordered (not all elements are comparable).

It turns out that **any set can be well-ordered**. It's just that the ordering may be arbitrary or not useful. It seems simple. Take any non-empty set S and define the ordering as follows. Choose an arbitrary element, say s, and make this the minimum element. Then take another element, different from s, say s', and make it the second, and so forth through all the elements of set S. The ordering so defined is a well-ordering, but of course *the ordering may have nothing to do with the 'usual' ordering* (if there is one). For example, an ordering of the rational numbers between 0.1 and 0.2 in *binary* representation is: 0.10, 0.11, 0.101, 0.111, 0.1001, 0.1101, 0.1011, 0.1111,...... (Exercise: show this is not the usual order.)

This argument depends on being able to arbitrarily choose an element from any set, even an infinitely large one. The mathematical justification for being able to do this is called the **axiom of choice**, which is assumed but not treated in this book.

Here is an ordering of all integers $\mathbb{Z}$ which is a well-ordering:
0, 1, 2, 3, 4, 5, ..., -1, -2, -3,......

Here is another well-ordering of all integers: 0, -1, 1, -2, 2, -3, 3,..... The symbols 0, 1, -1, 2,... are symbols and do not correspond to numbers with the usual ordering and arithmetic. Thus, 0 is first, -1 is second, 1 is third, -2 is fourth, 2 is fifth, etc (where usually there is no 'first integer' for all integers).

19 cardinal number manipulation

This is a relatively informal discussion of the concept of cardinality, but not an inaccurate one. The formalism may seem tedious, and it might be useful to first see a more informal discussion before encountering a more formal treatment (elsewhere).

We recall the notation: For set A,

card(A) = |A| = cardinality of A.
If A is finite, the cardinality is an integer, and equals the number of elements.
If A is countably infinite, card(A)=$\aleph_0$.
If A is uncountable and isomorphic to $\mathbb{R}$, card(A) = $\aleph_1$.

19.1 ordering cardinal numbers

For a finite set, the cardinal number of the set equals the number of elements of the set - aka size. Therefore to order cardinal numbers, we must order set sizes. We need to do this in a way such that we can compare the cardinal numbers of finite sets (ordinary numbers) and the cardinality of infinite sets ($\aleph_0$, $\aleph_1$). The basic idea is to define the comparison for finite sets such that the definition can be generalized to infinite sets as well.

Obviously we cannot compare the number of elements of sets directly, as that procedure is not defined for infinite sets.

Consider two finite sets A, B. Now consider an **injection**

$$F : A \to B \tag{19.1}$$

F is presumed injective but not necessarily surjective (ie, not necessarily bijective). We have seen that, for finite sets, the function F can be injective only when $|A| \leq |B|$ (section 7.7). This means that **an injection implies a method for comparing the cardinality |.| of two finite sets**. By moving from numbers to functions, we can transfer concepts of ordering from finite sets to infinite sets.

Definition 19.1

Given two nonempty sets A, B, (finite or infinite). Define the ***order relation*** $\preceq$ *such that* $|A| \preceq |B|$ *(card(A)* $\preceq$ *card(B)) when there exists some injection* $F : A \rightarrow B$.

$|\emptyset| \prec |X|$ *for X non-empty, and* $|\emptyset| \preceq |\emptyset|$.

The application of this concept to cardinality is as follows:

Consider two sets A, B and map F such that $F : A \rightarrow B$ is injective. If F is not bijective, then $|A| \prec |B|$. This is *interpreted* to mean that the cardinality of A is strictly less than the cardinality of B. For A, B finite, $\prec$ is the same as the usual $<$ such that $|A| \prec |B|$ means exactly the same thing as |A| < |B|. Consider two finite sets $A = \{1, 2, 3\}$, $B = \{1, 2, 3, 4, 5\}$. We can define an injection $F : A \rightarrow B$ as F(1)=1, F(2)=2, F(3)=3. There is no bijection and we **cannot** define an injection $F : B \rightarrow A$. This means $|A| \prec |B|$, which corresponds to our understanding that $3 = |A| < |B| = 5$.

On the other hand, letting $C = \{a, b, c\}$, there exists a bijection between A and C, so |A|= |C| =3.

For finite sets, **the implication of injection corresponds with our intuition about the cardinality (size) of one finite set being less than, greater than or equal to the cardinality of another finite set**. But we have moved from directly comparing numbers to using maps.

We can think of $\preceq$ as less than or equal to, and $\prec$ as strictly less than, but **these apply also to infinite sets**. We can loosely say that $|A| \preceq |B|$ means that card(A) is less than or equal to card(B), and if $|A| \prec |B|$ then card(A) is strictly less than card(B). But we cannot use the '$<$' or $\leq$ symbols in the usual way because card(A) and/or card(B) are not necessarily integers (they could be $\aleph_0$ or $\aleph_1$) and hence are not necessarily comparable. **The use of an injective function allows us to make the cardinality of any two sets, finite or infinite, comparable,** so the symbols $\prec$, $\preceq$ applies to finite and infinite sets.

In this way **we extend the comparison of cardinalities of finite sets to include infinite sets**.

Consider comparing set $A = \{1, 2, 3\}$ and $\mathbb{Z}_+$. Obviously there exists an injection $F : A \rightarrow \mathbb{Z}_+$ with F(1)=1, F(2)=2, F(3)=3. Since F is not bijective we have $|A| \prec |\mathbb{Z}_+|$. **The cardinality of A is strictly less than the cardinality of** $\mathbb{Z}_+$ $(3 \prec \aleph_0)$. Similarly there exists an injection $G : A \rightarrow \mathbb{R}_+$, where G(1)=1.1, G(2)=2.83, G(3) = 9804.62. Thus **the cardinality of A is strictly less than the cardinality of** $\mathbb{R}(3 \prec \aleph_1)$.

Theorem 19.2

Let A be any finite set and S any infinite set. Then $|A| \prec |S|$.

If card(A) = n, an integer, then this result means

$$n \prec \aleph_0 \qquad n \prec \aleph_1 \tag{19.2}$$

This is true for any finite integer n.

There is a bijection between any two countably infinite sets. Similarly there is a bijection between any two uncountable sets of cardinality $\aleph_1$.

Theorem 19.3

Let S, T be two countably infinite sets. Then

$$card(S) = card(T) \qquad \aleph_0 = \aleph_0 \tag{19.3}$$

Let X, Y be any two uncountable sets of cardinality $\aleph_1$. *Then*

$$card(X) = card(Y) \qquad \aleph_1 = \aleph_1 \tag{19.4}$$

Finally there is an injection (but not a bijection) from any any countably infinite set S into any uncountable set T (cardinality $\aleph_1$) and therefore

Theorem 19.4

Let S be countably infinite and T be uncountable. Then

$$card(S) \prec card(T) \qquad \aleph_0 \prec \aleph_1 \tag{19.5}$$

For this last case there is an injection from S into T but, as discussed previously, not a bijection, and therefore $\aleph_0 \prec \aleph_1$.

We have the following total ordering of cardinalities, where n is any (finite) integer

$$\begin{gathered} 1 \prec 2 \prec 3 ...... \prec \aleph_0 \prec \aleph_1 \\ n \preceq n \quad \aleph_0 \preceq \aleph_0 \quad \aleph_1 \preceq \aleph_1 \end{gathered} \tag{19.6}$$

Equations 19.6 corresponds to or intuition that any uncountable set ought to be larger than a countably infinite one, and any infinite set ought to be larger than a finite one, and now this intuition is quantitative.

19.2 addition of cardinal numbers

Now we define addition of cardinal numbers. So far we have defined the following cardinal numbers:

ordinary finite integers (cardinal number of finite set)
$\aleph_0$
$\aleph_1$

We want to define addition between cardinals of any two types, including the same type, so $n + \aleph_0$, $\aleph_0 + \aleph_1$ $\aleph_1 + \aleph_1$, and so forth. This means we need a symbol for addition of cardinality of sets and which (1) it reduces to ordinary + when both sets are finite and (2) it is a generalization of + which applies to addition involving infinite sets. We will use the symbol $\boxplus$ (called *boxplus*).

The plan is to define $\boxplus$ for finite sets in such a way that the definition is the same as for the usual + but also can be generalized to infinite sets.

First consider finite sets A, B. The *disjoint union* has the property that

$$|A \sqcup B| = |A| + |B| \tag{19.7}$$

with + the usual addition. This is the key relation. **We define addition of cardinality as the result of a disjoint union.** Because disjoint unions are associative and commutative, then so will be our definition of $\boxplus$. Here is the generalization:

Definition 19.5

For any two sets A, B (finite or infinite), we define ***cardinal number addition*** $\boxplus$ *as*

$$|A| \boxplus |B| = |A \sqcup B| \tag{19.8}$$

Recall that $|\emptyset| = card(\emptyset) = 0$.

If both sets A, B are finite, then $|A| \boxplus |B| = |A| + |B|$, but this is not the case when one or both sets are infinite since the symbol '+' is not defined in that case.

Consider $A = \{1, 2, 3\},\ \ B = \{1, 2, 3, 4\}$,
$|A| \boxplus |B| = |\{(0, 1), (0, 2), (0, 3)\} \cup \{(1, 1), (1, 2), (1, 3), (1, 4)\}| = 7.$

Then $|\mathbb{Z}_+| \boxplus |\mathbb{Z}_+| = |\{(0,1),(0,2),...\} \cup \{(1,1),(1,2),...\}|$. The right side, inside the $|...|$, is a finite union of two countably infinite sets and the result is countably infinite. so $|\mathbb{Z}_+| \boxplus |\mathbb{Z}_+| = \aleph_0$. This equation can be re-written as $\aleph_0 \boxplus \aleph_0 = \aleph_0$.

Continuing along these lines , using the above paragraph as a template, it is straightforward to show:

Theorem 19.6

The following is ***cardinal addition.*** *If x, y, z are cardinal numbers of finite sets (and so are ordinary integers) where $x + y = z$ where $+$ is ordinary addition*

$$
\begin{aligned}
& x \boxplus y = x + y = z \\
& x \boxplus \aleph_0 = \aleph_0, \qquad x \boxplus \aleph 1 = \aleph_1 \\
& \aleph_0 \boxplus \aleph_0 = \aleph_0, \qquad \aleph_1 \boxplus \aleph_1 = \aleph_1 \\
& \aleph_0 \boxplus \aleph_1 = \aleph_1
\end{aligned} \tag{19.9}
$$

The addition is commutative, so for example $x \boxplus \aleph_0 = \aleph_0 \boxplus x = \aleph_0$ and $\aleph_0 \boxplus \aleph_1 = \aleph_1 \boxplus \aleph_0 = \aleph_1$.

The disjoint union of a countably infinite set of integers and any interval of the real line is a union of a countably infinite set and an uncountable one. This union is uncountable, as discussed previously. This means exactly that $\aleph_0 \boxplus \aleph_1 = \aleph_1$. The other relations follow similarly. (Exercise: show this is you have doubts.)

Cardinal addition is also associative. For example

$$
\begin{aligned}
\aleph_0 \boxplus \aleph_0 \boxplus \aleph_0 = (\aleph_0 \boxplus \aleph_0) \boxplus \aleph_0 = \aleph_0 \boxplus \aleph_0 = \aleph_0 \\
\aleph_0 \boxplus \aleph_1 \boxplus \aleph_0 = (\aleph_0 \boxplus \aleph_1) \boxplus \aleph_0 = \aleph_1 \boxplus \aleph_0 = \aleph_1
\end{aligned} \tag{19.10}
$$

Multiplication and exponentiation of cardinals can also be defined, but we move on.

20 hierarchy of infinities

Having defined an order relation on cardinal numbers, we can now define a hierarchy of infinities - a sequence of infinite sets where the cardinality of each set in the sequence is infinite and strictly greater than the cardinality of the previous member of the sequence. It is not obvious that the procedure described below captures all possible infinite cardinalities, but at least it indicates an infinite number of successively greater (in sense of $\prec$ previously defined) infinite cardinalities.

It's all based on the following theorem

Theorem 20.1 *Given set S and power set $\mathcal{P}(S)$, then*

$$|S| \prec |\mathcal{P}(S)| \tag{20.1}$$

For any infinite S, $\mathcal{P}(S)$ is also infinite and any injection $F : S \to \mathcal{P}(S)$ is not a bijection.

This is intuitive, as each element of S is in its own singleton set in $\mathcal{P}(S)$, which also contains sets which are not singletons. Therefore no injection $F : S \to \mathcal{P}(S)$ can be a surjection, and there certainly exist injections (left as an exercise).

Let $S = \mathbb{Z}_+$. We have a hierarchy of infinitely large sets

$$|S| \prec |\mathcal{P}(S)| \prec |\mathcal{P}(\mathcal{P}(S))| \prec |\mathcal{P}(\mathcal{P}(\mathcal{P}(S)))| \prec .... \tag{20.2}$$

The cardinality of S is $\aleph_0$ but what is the cardinality of $\mathcal{P}(S)$? Is it $\aleph_1$, is there some infinity between $\aleph_0$ and $\aleph_1$, or is it some infinity greater than $\aleph_1$?

It turns out that *formal* set theory (which is beyond the scope of this book) can be consistent in either case, so that the resolution of this question must be a hypothesis rather than a theorem susceptible to proof. So we are led to the **continuum hypothesis: There is no cardinality between $\aleph_0$ and $\aleph_1$.** In other words, the continuum hypothesis is: $\aleph_1 = |\mathcal{P}(\mathbb{Z}_+)|$.

The notation for the cardinality of power sets is a generalization of equation 17.2. For any set W, finite or infinite, including any of the infinite sets introduced in this section,

$$|\mathcal{P}(W)| = 2^{|W|} \tag{20.3}$$

The continuum hypothesis is then

$$\aleph_1 = 2^{\aleph_0} \tag{20.4}$$

It turns out that one can define multiplication of cardinal numbers and exponentiation of a cardinal number (just as we defined addition of cardinal numbers), so the above two expressions are well defined, but this is not explained here.

21 ordinal numbers

21.1 introduction

The concept of ordinal numbers is a generalization of finite integers to what we could call (if we were so inclined) *integers which are infinitely large*, also called **transfinite numbers**. There are ordinal numbers which are finite and other ordinals which are infinite. The finite ones are the same as ordinary (finite) integers.

There can be a terminology problem here, as the usual concept of integer is of something which is finite. But infinite ordinal numbers have properties which are sufficiently similar to the properties of finite integers (but not identical) that we can loosely refer to them as *infinite integers* so long as we keep in mind what we are talking about.

Finite ordinal numbers are the same as finite cardinal numbers (ordinary integers), so unless we are dealing with infinitely large quantities, there is no difference.

A cardinal number is a generalization of the *size* of a set to infinite sets. As we have used the concept here, in an informal manner, a cardinal number was not described as a set (but see below). On the other hand, right from the start each ordinal number **is** exactly a set - it is a set of smaller ordinal numbers. An order relation between ordinal numbers is also an order relation among members of a set, as will be seen. This was not the case for cardinal numbers, where such an 'intra-set' order relation is not required (in the informal approach). Since a finite ordinal number is the same as an ordinary integer, each finite cardinal number is actually a finite ordinal number and hence must be *formally* handled as a set. We will not pursue this here.

The usual ordinal-cardinal differentiation states is that ordinals are used for **ordering** elements of a set while cardinals are used for counting the elements of a set, to determine the **size** of a set. That is, the ordinal numbers are first, second, third, ..., while cardinal numbers are 1,2,3,..., representing quantity. This way of looking at things may or may not be helpful in what follows.

Rigorous proofs of several theorems relevant to constructing ordinal numbers are beyond the scope of this book The general subject also covers way more material than is appropriate in a book of this nature. Therefore, this section is an intuitive and sufficiently precise introduction to some aspects of ordinal numbers, sufficient to develop some intuition. and familiarity with the terminology. The goal is to keep this section short and clear and far from exhaustive.

21.2 transitive sets

Transitive sets are tricky and it is crucial to understand the relationships of their elements. **Ordinal numbers are represented by transitive sets.**

Definition 21.1

A set S (finite or infinite) of sets is ***transitive*** *when, for any $s \in S$ also $s \subset S$. (However, $S \notin S$)*

The empty set $\emptyset$ is considered transitive.

A transitive set is a set of sets, since each element is also a subset. Examples of transitive sets all rely on the fact that the empty set $\emptyset$ is a subset of any set. So, for example, the set $S_1 = \{\emptyset\}$ is transitive, since $\emptyset$ is both an element of S_1 (the only element) and is also a subset of S_1 ($\emptyset \subset S_1$).

The set $S_2 = \{\,\emptyset, \{\emptyset\}\,\}$ is also transitive. Note that $\{\emptyset\} \subset S_2$ because the set $\{\emptyset\}$ is a set containing the 'first' element $\emptyset$. See the discussion below equation 5.39 in section 5.6.

We can keep going. The set $S_3 = \{\,\emptyset, \{\emptyset\}, \{\emptyset, \{\emptyset\}\}\,\}$ is transitive. The 'first two' elements are as above, and the 'third' element is a subset of S_3 (it is the subset containing the first two elements of S_3).

Exercise: Write the next largest transitive set.

The following countably infinite set is transitive.

$$\{\,\emptyset, \{\emptyset\}, \{\{\emptyset\}\}, \{\{\{\emptyset\}\}\}, ...\} \tag{21.1}$$

Theorem 21.2

Let A be a non-empty transitive set and X, Y non-empty sets. If $Y \in X \in A$, then $Y \in A$ and $Y \subset A$.

$Y \in X$ does not in general imply $Y \in A$. But since A is transitive, and since $X \subset A$, and since $Y \in X$, every element of X is in A, so then $Y \in A$. Since A is assumed transitive, then $Y \in A$ implies $Y \subset A$. This theorem is certainly not true for any arbitrary set.

'Transitive set' is a rather exclusive category of sets. For example, suppose S is a transitive set. Can S be of the form $S = \{A\}$, where A is some **nonempty** set? Then A must be such that $A \subset S$ which is not true for A non-empty because the only two subsets of S are $\{A\}$ and $\emptyset$. Therefore A cannot be nonempty. The only alternative, if S is to be a singleton, is $A = \emptyset$. Then $S = \{\emptyset\}$ is transitive because $\emptyset$ is a subset of S.

Can a finite transitive S be of the form $S = \{A, B\}$, where A and B are non-empty sets. The argument above shows that both A and B cannot be nonempty. However, this form of S is valid if A is the empty set and $B = \{A\}$, the form considered previously.

In other words, the so-called **edge condition**, theorem 4.9, is essential if we are to form transitive sets. And ordinal numbers are transitive sets.

21.3 defining finite ordinals - Von Neumann ordinals

Let's get a concrete idea of how a finite ordinal can be defined as a set, then consider some properties of those sets, and then consider infinite ordinals as an extrapolation of the finite ordinals. Some of the definitions and theorems of this section will also apply to infinite ordinals.

This representation of ordinal numbers is due to the mathematician John Von Neumann. The ordinal numbers are represented by the **symbols** 0,1,2,... and are **defined** as follows:

Define 0 (zero) as the empty set.

$$0 = \emptyset \tag{21.2}$$

We may now define 1 as the set containing 0 as its only element (and so the set 1 has one element)

$$1 = \{\emptyset\} = \{0\} \tag{21.3}$$

The set representing 2 is now the set with two elements, the elements being 0 and 1.

$$2 = \{\,\emptyset, \{\emptyset\}\,\} = \{0, 1\} \tag{21.4}$$

We may continue this process, though the explicit representation as sets involving the empty set starts to get hard to read.

$$3 = \{\ \emptyset, \{\emptyset\}, \{\,\emptyset, \{\emptyset\}\,\}\ \} = \{0, 1, 2\} \tag{21.5}$$

$$4 = \{0, 1, 2, 3\} \tag{21.6}$$

and so forth, always keeping in mind that **the symbols 0,1,2,3,... are actually sets**. (Exercise: write '4' in terms of the empty set.)

The ordinals defined above are sets have the following property:

Theorem 21.3

The ordinals 0,1,2,3,.... (as defined above) are each transitive sets. An ordinal is a transitive set of ordinals.

The reasoning is similar to the discussion of the previous section.

Any ordinal n (assume n $>$ 0) is of the form $n = \{0, 1, 2, ..., (n-1)\}$, where each of 0,1,2,...(n-1) are sets. Note that $0 \in n$ because it is there explicitly. But also $0 \subset n$ because $0 = \emptyset$, and $\emptyset$ is a subset of any set.

The ordinal $1 = \{0\} = \{\emptyset\}$ is an ordinal based on the above discussion.

Now consider $2 = \{0, 1\}$. The case of 0 is handled above. $1 \subset 2$ because $1 = \{0\} \subset \{0, 1\} = 2$. Note that $\{1\} \subset \{0, 1\}$ is not the same as $1 \subset \{0, 1\}$. The latter is valid because $1 = \{0\}$. (The former is true but not relevant.) Therefore $2 = \{0, 1\}$ is an ordinal.

Next consider $3 = \{0, 1, 2\}$. Both 0 and 1 are subsets of the set 3, as shown above. Then $2 = \{0, 1\} \subset 3 = \{0, 1, 2\}$.

It is seen that the subset relation depends crucially on an ordinal being defined as a set.

The set $A = \{5\} = \{\{0, 1, 2, 3, 4\}\}$ is not transitive. While $5 \in A$ (and 5 itself is transitive), it is the case that
$5 = \{0, 1, 2, 3, 4\} \not\subset A = \{\{0, 1, 2, 3, 4\}\}$. Rather $\{5\} = \{\{0, 1, 2, 3, 4\}\} \subseteq A$.

Theorem 21.4

The ordinals (finite and infinite) are ***totally ordered based on set inclusion*** $\in$. *That is , for ordinals x, y, $x \neq y$:*

$$x < y \iff x \in y \tag{21.7}$$

For finite ordinals, the ordering is the usual $\leq$ for integers.

That is, $1 < 2$ because $1 \in 2$, that is, $1 \in \{0, 1\}$. Then $8 < 20$ because
$8 \in 20 = \{0, 1, 2, 3, 4, 5, 6, 7, 8, ..., 19\}$.

Having defined a total order on ordinals, we have the following.

Definition 21.5

An ordinal number n (finite or infinite) is the set of all ordinals less than n, including 0. This set is transitive.

For now, consider finite ordinals. So for example the ordinal 3 is the set of ordinals $\{0, 1, 2\}$. The ordinal 108 is the set of ordinals $\{0, 1, 2, ..., 107\}$.

The set corresponding to any ordinal is well-ordered.

Theorem 21.6

The sets 0,1,2,3,... are each well-ordered (section 18.6).

Each set is totally ordered using the set-inclusive ordering defined above, each set has a minimum element (namely 0) and any subset has a minimum element. For example, $\{4, 6, 9\} \subset 10$ has a minimum element 4.

A set of ordinals is not necessarily an ordinal. For example, the set $\{2, 5, 8\}$ is not an ordinal because it is not transitive. However, $\bigcup\{2, 5, 8\} = 8$ is an ordinal. If S is a set of ordinals (not necessarily transitive and hence not necessarily an ordinal) then $\bigcup S$ is an ordinal. In fact, if S has a largest element, the union must equal the largest element of S. This is certainly obvious for finite ordinals (write a few out to convince yourself). We will see that it applies also to certain infinite ordinals.

21.4 addition of ordinals - an example

One thing we must consider is whether the finite numbers defined by equations 21.2, 21.3, 21.4, etc can be made to obey the usual laws of arithmetic. To continue, we must at least consider addition of ordinals, when the ordinals are defined as above. We want to define addition + of ordinals (which are defined as above) for finite and infinite ordinals. There are several ways to do this, but I think the clearest and simplest to work with uses the concepts of order type, order isomorphism and disjoint union.

To appreciate what is involved, consider adding the ordinals 2 and 3. How do we define $2 + 3$? Recall $2 = \{0, 1\}$, $3 = \{0, 1, 2\}$. Using the same trick that was used for addition of cardinals, **define**

$$2 \boxplus 3 = 2 \sqcup 3 = (\{0\} \times 2) \cup (\{1\} \times 3) = \{(0, 0), (0, 1), (1, 0), (1, 1), (1, 2)\} \tag{21.8}$$

The set on the right is ordered in the dictionary order. Thus $(0,n) < (1,n')$ for any $n, n' \geq 0$. Also, for k = 0 or 1, $(k,n) < (k,n')$ when $n < n'$.

We now relate the set on the right of equation 21.8 to the ordinal 5. The trick is that there is a bijection between the set on the right side of equation 21.8 and $5 = \{0,1,2,3,4\}$:

$$(0,0) \to 0, (0,1) \to 1, (1,0) \to 2, (1,1) \to 3, (1,2) \to 4 \tag{21.9}$$

Therefore

$$2 \boxplus 3 \cong 5 \tag{21.10}$$

which is an isomorphism. This bijection **preserves order**. Thus $(0,0) < (1,0)$ and $0 < 2$. $(1,1) < (1,2)$ and $3 < 4$. An isomorphism such as this which preserves order is called an *order isomorphism* and is discussed in the next section.

The next trick is to replace $\cong$ by $=$, to get $2 \boxplus 3 = 5$, and then replace $\boxplus$ by $+$ to get $2 + 3 = 5$. This is just the replacement of one symbol by an overloaded one.

The addition of ordinals is represented by a disjoint union followed by an isomorphism. The generalization of this procedure is valid for finite ordinals, for which we get the usual result. But it is also valid for infinite ordinals (ie, for infinite well-ordered transitive sets of ordinals). But first we need the concept of order isomorphism and order type.

21.5 order type and order isomorphism

Definition 21.7

Given S, T well-ordered sets. If there exists some bijection $F : S \to T$ such that, for $s \in S$, $t \in T$,

$$s \leq t \Longleftrightarrow F(s) \leq F(t) \tag{21.11}$$

the sets are said to be ***order isomorphic****. We write*

$$S \cong_o T \tag{21.12}$$

The subscript 'o' on $\cong$ indicates an order isomorphism.

The bijection preserves the well-ordering of the sets. If two sets are isomorphic and well-ordered, then order isomorphism works as follows. For each set, select the minimum element (which must exist as sets are well-ordered). These are paired in the bijection. Then choose the next smallest element and pair these, and so forth. Since the sets are isomorphic, this procedure must pair all the elements of each set in proper order.

An important point about order isomorphism is the following. **If two sets are order isomorphic (with given order relations) then the bijection is unique.** Obviously the bijection will depend on the order relations of the two sets, but the discussion above indicates that there is only one way to create a bijection which preserves order.

The following theorem (which is reasonable but which will not be proven) is the heart of the matter.

Theorem 21.8

Any well-ordered set (finite or infinite) is order-isomorphic to one and only one ordinal.

For example, the isomorphism, equation 21.9, is an order-isomorphism between the sets $2 \sqcup 3$ and the ordinal 5. The alphanumeric set $\{a, b, c, d, e, f, g\}$, with alphanumeric order a < b < c< ... is order isomorphic to ordinal 7. This theorem is self-evident for finite sets (and finite ordinals), and it is also valid for infinite sets and infinite ordinals. *The well-ordered set need not be transitive nor an ordinal.* An order isomorphism does not preserve 'transitiveness' or 'ordinalness' but only order.

Since any well-ordered set is order isomorphic to exactly one ordinal, we have:

Definition 21.9

Given a well-ordered set S order isomorphic to ordinal x. We say that S has the ***order type*** *of x.*

Therefore

Definition 21.10

If two well-ordered sets A, B are order isomorphic to the same ordinal x, they are order isomorphic to each other and both have order type x.

$$A \cong_o B \cong_o x \tag{21.13}$$

21.6 ordinal addition

Definition 21.11

ordinal addition. *Given ordinals x, y (finite or infinite) whose the sum is ordinal z. The addition is notated*

$$z = x + y \tag{21.14}$$

where z is defined as the ordinal order isomorphic to $x \sqcup y = (\{0\} \times x) \cup (\{1\} \times y)$ and '$=$' is defined as $\cong_O$. The ordering is important.

The set $x \sqcup y$ is ordered in the dictionary order. Thus $(0, n) < (1, n')$ for any $n, n' \geq 0$. Also, for k = 0 or 1, $(k, n) < (k, n')$ when $n < n'$.

This is an overloading of $=$.

This exactly reproduces the usual arithmetic addition for finite numbers, and it can be extended to apply to infinite ordinals. The previous example, section 21.4, illustrates this definition. Here is another example. Consider $z = 12 + 7$.

1. Define $S = 12 \sqcup 7$. $S = \{\ (0,0), (0,1), , ..., (0,11),\ (1,0), (1,1), ..., (1,6)\ \}$.

2. This is order isomorphic to $19 = \{0, 1, 2, ..., 18\}$, where
$(0,0) \to 0, (0,1) \to 1, ..., (0,11) \to 12,\ (1,0) \to 13, (1,1) \to 14, ..., (1,6) \to 18$.

3. Therefore $12 + 7 = \{0, 1, 2, ..., 18\} = 19$.

It should be obvious that addition of **finite** ordinals is both commutative and associative.

However, when adding infinite ordinals, the symbol '$=$' does *not* behave exactly as it does for finite ordinals (in that sense it is an abuse of notation). In particular, we will see that one consequence is that addition may not be commutative when it involves infinite ordinals. One must keep in mind this special meaning of the symbol '$=$' when adding infinite ordinals.

21.7 infinite ordinals

For infinite ordinals, the following definition is useful.

Definition 21.12

*Let n be an ordinal (finite or infinite). Then the ordinal $n+1$ is called the **successor** of n and is written.*

$$n + 1 = succ(n) \cong_o n \sqcup \{1\} \tag{21.15}$$

As will be seen, $succ(n) \neq 1 + n$ unless n is finite.

Here is an important property of successors.

Theorem 21.13

If ordinal n contains a maximum element (say ordinal N), then $n = succ(N)$.

For example, $8 = \{0, 1, 2, 3, 4, 5, 6, 7\}$ and $8 = \text{succ}(7)$. We will see that **every ordinal has a successor, but not every infinite ordinal is a successor of some other ordinal.** This property is a significant difference between finite and infinite ordinals.

For finite ordinals, as shown in the previous section, $n+1 = 1+n$, but, as will be seen, this is not the case for an infinite ordinal.

If $\mathbb{N}$ is the set of natural numbers (positive integers and 0), then the symbol ω (omega) is defined as

$$\omega = \mathbb{N} = \{0, 1, 2, 3, ...\} \tag{21.16}$$

This is common notation. An important point is: **every element of ω is a finite ordinal.** Now we define an infinite ordinal.

Theorem 21.14

$\omega = \{0, 1, 2, 3, ...\}$ is an infinite ordinal.

ω is greater than any finite ordinal.

ω is the smallest infinite ordinal.

The ordinal ω is not a successor of any ordinal.

ω has no maximum element, but 0 is its minimum element.

In other words, ω is the smallest ordinal which is larger than any finite number.

The set ω is an infinite set, so if ω is an ordinal, it is an infinite ordinal. It is an ordinal because, for any $n \in \omega$, $((n = \{0, 1, 2, ..., n-1\})$ a finite ordinal) $n \subset \omega$ because n is the set of all ordinals less than n and they are in ω. Therefore ω is a transitive set of ordinals and so is an ordinal.

For any finite ordinal n, $n \in \omega$ and therefore $\omega > n$ (theorem 21.4, order relation of ordinals), so ω is greater than any finite ordinal.

If ω were not the smallest infinite ordinal, then because all ordinals are ordered, there must be some infinite ordinal $\alpha < \omega$ which means $\alpha \in \omega$, but the set ω contains only finite ordinals. Therefore ω is the smallest infinite ordinal.

ω is not a successor of any finite ordinal. Suppose there were some $n \in \omega$ such that $\omega = n+1$. But $n+1$ is a finite ordinal and ω is infinite. This is a different way of saying that ω has no maximum element, because if it did, then ω would be its successor (theorem 21.13).

We point out that the collection of all ordinals is not a set, because such a set is an ordinal and hence a member of itself and therefore is not a valid set.

Definition 21.15

An ordinal which is not a successor (and hence has no maximum element) is called a ***limit ordinal****.*
ω ***is a limit ordinal.***

Limit ordinals are infinite well-ordered transitive sets. But not every well-ordered infinite transitive set is a limit ordinal, as shown below.

Now consider the set

$$W = \{0, 1, 2, 3, ..., \omega\} \tag{21.17}$$

This is an ordinal - a well-ordered transitive set of ordinals - and ω is its maximum element (ω is larger than any finite ordinal). Therefore W is not a limit ordinal. In addition, we now show

$$W = \omega + 1 = succ(\omega) \tag{21.18}$$

Extending the addition of ordinals to infinite ordinals

$$\omega + 1 \cong_o (\{0\} \times \omega) \cup \{\,(1, \omega)\,\} = \{(0, 0), (0, 1), (0, 2), ..., (1, \omega)\} \tag{21.19}$$

Note that $W \neq \omega$ because $w \notin \omega$ (ω does not contain itself) but $\omega \in W$. Also $W > \omega$ because of the definition of ordering of ordinals.

Note that **the ordering of ordinals is not in general described by the existence of an injection** as it is for cardinals. There is a bijection $w \to W$, since $|\omega| = |W|$. Then $\omega + 1$ is not order-isomorphic to ω. It is larger than ω ($\omega \prec \omega + 1$ because $\omega \in \omega + 1$)

We now show

$$\omega + 1 \neq 1 + \omega \tag{21.20}$$

because $1 + \omega = \omega$. That is, $1 + \omega$ is order isomorphic to ω, where the ordering is as in definition 21.11.

$$\begin{aligned} &1 + \omega = \{(0, 1), (1, 0), (1, 1), (1, 2), ... \\ &(0, 1) \to 0 < (1, 0) \to 1 < (1, 1) \to 2, ... \end{aligned} \tag{21.21}$$

> Addition of two ordinals is not in general commutative if at least one of them is an infinite ordinal. However, addition is associative.

If two ordinals are not order isomorphic, they are not equal.

Index

Made in the USA
Middletown, DE
05 November 2026

36588048R00106